# CARS AND CULTURE

# CARS AND CULTURE

## THE LIFE STORY
## OF A TECHNOLOGY

Rudi Volti

**The Johns Hopkins University Press**
Baltimore

© 2004 by Rudi Volti

Printed in the United States of America on acid-free paper

First Published in 2004 by Greenwood Press

Johns Hopkins Paperbacks edition published 2006

9  8  7  6  5  4  3  2  1

The Johns Hopkins University Press

2715 North Charles Street

Baltimore, Maryland 21218-4363

www.press.jhu.edu

ISBN 0-8018-8399-7 (pbk.: alk. paper)

**Library of Congress Control Number: 2005936768**

A catalog record for this book is available from the British Library.

Published in an edited paperback edition by arrangement with Greenwood Publishing Group, Inc., Westport, CT.

# Contents

# Acknowledgments

I am indebted to the following people who have taken the time to read individual chapters of the manuscript and to call my attention to errors of fact or interpretation: Ann Stromberg, Mark Rose, and Robert Post. Any remaining errors are, of course, my exclusive responsibility.

# Introduction

More than any other artifact of modern technology, the automobile has shaped our physical environment, social relations, economy, and culture. At the same time, however, the automobile has not come into our lives as an alien force. Although our embrace of the automobile is often accompanied by unease over many of its consequences, it cannot be denied that the ownership and operation of cars resonate with some of our most important values and aspirations. We have paid a high price for the personal mobility that the car offers, but for the most part we have done so willingly, if not always intelligently.

A complete assessment of the significance of the automobile would require a library of many volumes, and some key works in the history of the automobile are presented in the Bibliography. Here, our aims are necessarily limited. To put it as succinctly as possible, this book is intended to provide a brief introduction to the history of the automobile, paying particular attention to the automobile's technical evolution while at the same time delineating the cultural, social, and political context in which that evolution has taken place. Space limitations often have prevented an exhaustive exploration of many important topics, but the reader of this book should be well prepared to take on more specialized studies of the automobile, its impact, and the forces that have shaped its development.

The technical development of the automobile and its constituent parts have been sources of endless fascination, and one of the prime goals of this book is to narrate how the primitive horseless carriages of the late nineteenth century were transformed into the fast, comfortable, and reliable vehicles that we take for granted today. But as important as the development of automotive technology has been, this book is about more than the evolution of engines, transmissions, suspensions, accessories, and bodywork. Along with being a technological creation, the car has been the product of myriad nontechnical forces. Although the automobile provides an individualistic and privatized approach to transportation, it also has required collective efforts on a massive scale. The manufacture of cars by the millions has been the work of gigantic industrial firms, and any comprehensive history of the automobile has to take into consideration the development of large-scale production technologies, the rise and frequent fall of particular firms, and the accomplishments and failures of the entrepreneurs and managers who created and administered these enterprises.

Another key element of the collective activity essential to automobile-based transportation is government. Although automobile ownership has been most extensive in countries with capitalist, market-driven economies, governments have played an indispensable role in providing the infrastructure essential to automobile operation. Cars are of little value if they do not have adequate roads to travel on, and road construction has been an activity best done by some level of government. For many decades roads and the enforcement of traffic regulations were the chief areas of government involvement in automotive matters, but this began to change in the 1960s when the dangerous and destructive consequences of widespread automobile ownership became impossible to ignore. Concerns about safety, air pollution, and energy use have been reflected in a larger role for government regulation of the automobile, and efforts in this area will receive a fair amount of emphasis here.

While many business firms have made large profits from it and governments have attempted to both support and control it, the automobile has brought a multitude of social and cultural changes everywhere it has gained a toehold. Adequate coverage of the history of the automobile, therefore, requires that attention be given to drive-in restaurants, labor unions, domestic architecture, gender relations, and the growth of suburbia. All of these topics, along with other examples of the automobile's influence on the world we live in, will be examined in the pages that follow.

Although the automobile was not an American invention, the United States was the first nation to experience mass motorization. Not only was the ratio of cars to people far greater in the United States than in any other nation for many decades, but also many of the key technologies and organizational methods that made mass motoring possible originated in the

United States. Consequently, much of the content of this book centers on developments that have taken place on American soil. At the same time, however, an effort also has been made to examine in some detail how an automotive-related culture has evolved in Europe and other parts of the world. While these experiences are important in their own right, they also help us to see how the development of a particular technology has been strongly influenced by economics, culture, and government policies.

## THE PLAN OF THE BOOK

The basic organizational format of this book is chronological. Each chapter focuses on a period of time delineated by a key phase in the history of the automobile, or by historical epochs like the economic depression of the 1930s or the boom years that followed World War II. But events taking place in the real world do not always fit into neat historical compartments, and at times it has been necessary to take the narrative beyond the chronological confines of the individual chapters in order to tell a coherent story. The need to deviate from strict chronological order also manifests itself in occasional repetitions of things said in previous chapters.

As a final point, any discussion of technical features and their development entails some use of a specialized vocabulary. Although jargon has been kept to a minimum, some use of technical terms is unavoidable. Efforts have been made to explain these in the body of the text and in the glossary that appears at the end of the book.

# Timeline

| | |
|---|---|
| 1765–1770 | Nicholas Cugnot builds two steam-powered vehicles intended to haul artillery. |
| circa 1830 | Steam coaches in regular service in England. |
| 1839 | William Grove invents the fuel cell. |
| 1862 | Etienne Lenoir builds a vehicle powered by an internal combustion engine of his own design. |
| 1865 | Britain enacts the Locomotives on Highways Act, commonly known as the Red Flag Act. |
| 1876 | Nicolaus Otto invents the four-stroke internal combustion engine. |
| 1885 | Gottlieb Daimler and Wilhelm Maybach combine an internal combustion engine with a wooden-framed, two-wheeled vehicle. |
| | Carl Benz builds a three-wheeled vehicle with an internal combustion engine. |
| 1886 | Daimler and Maybach construct a four-wheeled vehicle with an internal combustion engine. |
| 1888 | The pneumatic tire is reinvented by John B. Dunlop. |
| 1891 | The firm of Panhard and Lavassor introduces an automobile with a front-mounted engine. |

1892    The National League for Good Roads established in the United States.

1893    Charles and Frank Duryea create the first American automobile powered by an internal combustion engine.

1894    The first automobile competition takes place in France.

Wilhelm Maybach invents the spray carburetor.

1895    The first automobile race in the United States begins in Chicago.

Patent for the automobile issued to George B. Selden.

1896    Henry Ford builds his first automobile.

1899    An electric car becomes the first automobile to reach 60 mph.

1900    Ferdinand Porsche creates a car that uses both an internal combustion engine and electric motors.

1901    The firm headed by Ransom E. Olds sells 425 Oldsmobiles.

New York becomes the first state to require the registration of automobiles.

1903    H. Nelson Jackson and Sewall Crocker make the first automobile trip across the United States.

The Ford Motor Company is established in Detroit.

1904    The United States passes France to become the world's largest manufacturer of automobiles.

1905    The first drive-in gas station in the United States opens in St. Louis.

1906    A steam-powered Stanley achieves 127.7 mph at Ormond Beach, Florida.

Speed limits on open highways are in force in most states.

The aftermath of the San Francisco earthquake demonstrates the utility and reliability of automobiles.

1908    The Ford Model T is introduced.

General Motors is incorporated.

1910    William Durant ousted as head of GM.

1911    Ford sets up operations in England.

1912    All-metal bodies used for some models by Hupmobile and Oakland.

Cadillac equips its cars with electric starters.

| | |
|---|---|
| 1913 | Ford sets up the first assembly line, which is used to manufacture a magneto component used by the Model T. |
| | Thermal cracking doubles the amount of gasoline that can be refined from a given quantity of petroleum. |
| 1914 | Ford introduces the $5-per-day wage. |
| | World War I erupts; Parisian taxicabs ferry French troops to the front line. |
| | Dodge markets the first mass-produced cars with all-metal bodies. |
| 1915 | William Morris begins to manufacture Britain's first mass-produced car, the Cowley. |
| 1916 | William Durant regains control of General Motors. |
| | The Federal Aid Road Act is enacted in the United States, providing federal matching funds for road construction. |
| 1919 | Oregon, New Mexico, and Colorado become the first states to tax gasoline. |
| | Citroën begins to use an assembly line. |
| 1920 | More than 30 percent of American farms have an automobile. |
| | Alfred Sloan takes over the presidency of General Motors. |
| 1921 | Britain enacts a horsepower tax. |
| | Essex introduces the first popularly priced enclosed car. |
| | The Pig Stand drive-in restaurant opens in Texas. |
| 1922 | Lancia produces the first car using unit construction. |
| 1923 | More than 1,800,000 Model Ts are sold by the Ford Motor Company. |
| | General Motors introduces the annual model year. |
| | Walter Chrysler becomes head of the Maxwell Motor Company. |
| | Leaded gasoline begins to be sold. |
| 1924 | The Chrysler Six is introduced. |
| 1925 | The first Italian autostrada is opened. |
| | General Motors acquires Vauxhall. |
| | Sales of closed cars exceed those of open cars in the United States. |
| | The Motel Inn opens in San Luis Obispo, California. |

1927    The last Model T is produced.

General Motors establishes the Art and Colour Section.

1928    Sales of the Ford Model A begins.

Coast-to-coast bus service is inaugurated in the United States.

Chrysler introduces the low-priced Plymouth.

1929    General Motors acquires Opel.

Synchromesh transmission used on some General Motors cars.

The stock market crash heralds the Great Depression
of the 1930s.

1930    Ford participates in the construction of a large automobile plant in
the Soviet Union.

1931    The first Datson (later Datsun) is produced in Japan.

1932    Ford introduces a low-priced, V-8 powered car.

1933    The Nazi Party comes to power in Germany and expands the construction of autobahns.

The first drive-in movie theatre is opened in Camden, New Jersey.

1934    The Chrysler Airflow is produced.

1935    The Opel Olympia becomes the first mass-produced car using unit
construction.

1936    Mercedes offers a diesel-powered car, the 260D.

Three prototype Volkswagens are delivered to the German
government.

1937    Sitdown strikes begin at General Motors.

1938    Construction begins on the Pennsylvania Turnpike.

1939    The Hydra-Matic transmission is available on Oldsmobile models.

The General Motors Futurama exhibit opens at the New York
World's Fair.

1940    The Arroyo Seco Parkway opens in Southern California.

The prototype Jeep is exhibited at Fort Benning, Georgia.

Air conditioning available on a few Packards.

1942    Production of automobiles for civilian use ceases in the United
States.

| | |
|---|---|
| 1944 | The Federal Aid Highway Act of 1944 is signed by President Roosevelt. |
| 1945 | Henry Ford retires as head of the Ford Motor Company, and his place is taken by his grandson, Henry II. |
| 1947 | Michelin introduces the radial tire. |
| 1948 | Cadillac and Oldsmobile introduce modern overhead valve V-8 engines for the 1949 model year. |
| 1950 | Chrysler introduces the "Hemi" engine for the 1951 model year. |
| 1952 | The first Holiday Inn opens in Memphis, Tennessee. |
| 1954 | Studebaker and Packard merge to form Studebaker-Packard. |
| | Nash and Hudson merge to form American Motors. |
| 1955 | Germany surpasses Britain as the world's largest exporter of automobiles. |
| | Ford offers a number of safety features for the 1956 model year. |
| | The first franchised McDonald's restaurant opens in Des Plaines, Illinois. |
| 1956 | President Dwight D. Eisenhower signs the Interstate Highway Act. |
| 1957 | Ford brings out the Edsel for the 1958 model year. |
| 1959 | British Motor Corporation introduces the Mini. |
| 1964 | The Ford Mustang is introduced as a 1965 model. |
| 1965 | *Unsafe at Any Speed* by Ralph Nader is published. |
| | The federal government enacts the Motor Vehicle and Air Pollution Act. |
| 1966 | The U.S. Congress passes the National Traffic and Motor Vehicle Safety Act. |
| | The California legislature sets emission standards higher than those set by the federal government. |
| 1968 | Cars sold in the United States are required to have seat belts. |
| 1970 | Congress passes the Clean Air Act and creates the Environmental Protection Agency to enforce emissions standards. |
| | Amtrak takes over money-losing passenger trains from privately owned railroads. |
| 1971 | Japan becomes the world's second-largest producer of automobiles. |

| | |
|---|---|
| 1972 | Workers at General Motors' Lordstown plant go on strike to protest high assembly-line speeds. |
| 1973 | The U.S. Congress allows a portion of the Interstate Highway Trust Fund to be used for public transit. |
| 1973–1974 | The first energy crisis is sparked by the Yom Kippur War. |
| 1975 | The Energy Policy and Conservation Act mandates Corporate average fuel economy (CAFE) standards. |
| 1979 | Another gasoline shortage follows the Iranian revolution. |
| | General Motors introduces a line of front-wheel drive cars for the 1980 model year. |
| | The federal government agrees to provide loan guarantees for Chrysler. |
| 1981 | The Reagan administration negotiates "orderly marketing arrangements" that limit the export of Japanese cars to the United States. |
| 1982 | Honda begins producing cars in Marysville, Ohio. |
| 1984 | Chrysler minivans go on the market. |
| 1985 | General Motors and Toyota begin joint production in a former GM plant in Fremont, California. |
| 1995 | The number of cars and light trucks in the United States equals the number of licensed drivers. |
| 1998 | Chrysler and Daimler-Benz merge to form DaimlerChrysler. |
| 2001 | Sales of SUVs, pickup trucks, and minivans equal those of conventional passenger cars. |
| 2003 | General Motors recalls its fleet of leased electric EV1s. |
| | The number of cars and light trucks in the average American family unit exceeds the number of licensed drivers in the household. |

# 1

# The Dawn of the Motor Age, 1886–1905

◆

During the last decade and a half of the nineteenth century, a number of crude self-powered devices clattered out of workshops in Germany and France. Although it may not have been apparent at the time, the progeny of these early automobiles were destined to have a profound impact on the world. During the early years of the succeeding century, the automobile made rapid strides in speed, reliability, and practicality, although many problems remained to be solved. Undaunted by technical and financial problems, thousands of hopeful entrepreneurs went into the business of manufacturing automobiles. The great majority failed, but those that succeeded were assured of a vast market for their products. Although cars and their drivers initially were reviled by many, the automobile's allure proved irresistible, and the ranks of car owners grew at a rapid pace. The mushrooming popularity of the automobile prodded national and local governments to try to keep the automobile under their control, an endeavor in which they met with only partial success.

## THE EMERGENCE OF THE AUTOMOBILE

When was the first automobile built, and who built it? Before answering these questions we must be mindful that the word "first" always has to be used with caution when recounting the history of a particular technology. To begin, it is necessary to precisely define the invention in question, so in describing the origins of the automobile we need to make clear what is meant by "automobile." If we take the most general definition—a self-propelled vehicle running on a road surface—credit for its invention usually goes to Nicholas Cugnot (1725–1804), who from 1765 to 1770 built two steam-powered vehicles that were intended to haul artillery for the French army. The first of Cugnot's vehicles was said to travel steadily at 3 mph (5 kph), but ran out of steam after 15 minutes. His second showed some promise, but after it ran into a wall on a test run, the government lost interest and ended financial support for the project.

Steam continued to be the logical source of power for much of the century that followed. Roughly paralleling the development of railroad locomotives, steam-powered tractors, wagons, carriages, and even bicycles were created by inventors in several parts of the world. Beginning around 1830, steam coaches enjoyed some commercial success in England. These might have paved the way for individually driven steam-powered automobiles, but the rapid growth of railroad systems (which were not encumbered by the poor roads of the time) and opposition from existing stagecoach lines resulted in their premature demise.

Although these vehicles demonstrated that a nonorganic power source could be used for propulsion, the invention that had the most immediate influence on the early history of the automobile was the bicycle. Although it relied on human muscle power, the bicycle was important for two reasons. First, many of the material and production technologies that came to be used for automobiles originated with the bicycle. Chain-and-sprocket drive, pneumatic tires, ball bearings, tension-spoke wheels, advanced metal-stamping techniques, and brazed tube construction all owed a great deal to the bicycle industry that boomed in the 1880s. Equally important, if more difficult to precisely demarcate, was the bicycle's promotion of a culture of personal mobility. Toward the end of the nineteenth century, rail and ship transportation allowed people to travel great distances in relative comfort, but only in accordance with the schedules set by railroad and ship companies. In contrast, bicycles allowed their riders to go anywhere there was a road to take them, at any time they wanted. The importance of the bicycle was noted by Hiram Maxim, a pioneering automobile manufacturer, who answered his rhetorical question as to why the automobile had not emerged prior to the 1880s by noting that "the bicycle had not yet come in numbers and had not directed men's minds to the possibilities of independent, long-distance travel over the ordinary highways" (Maxim 1937, 5).

## THE INTERNAL COMBUSTION ENGINE

Influential though it was, the idea of personal mobility would have not been widely realized without an effective source of power for individual vehicles. Steam engines were a proven technology, but as we shall see they were not well suited to automobiles. A better proposition was the internal combustion engine, so called because the combustion of an air-fuel mixture occurs inside the engine (in contrast, a steam engine uses external combustion to heat the water in a boiler). Efforts to build internal combustion engines go back to the seventeenth century when a few inventors tried to construct engines fueled by gunpowder. A bit more practical was a handful of engines built in the mid-nineteenth century, most notably the ones constructed by Etienne Lenoir (1822–1900) in France. These engines were double acting; an ignited mixture of air and illuminating gas pushed a piston from one end of a cylinder to the other. The process then was repeated, with the piston producing another power stroke as it moved back, while at the same time pushing out the exhaust gases left over from the first power stroke. This meant that two power strokes were produced for each revolution of the crankshaft. The weakness of this design was that the air-fuel mixture was not compressed prior to combustion, resulting in limited power and efficiency. Even so, Lenoir's engine had a higher thermal efficiency than steam engines of equivalent size. Lenoir was even able to use one of his engines to power a vehicle in 1862, but as it reportedly required six hours to cover a distance of six miles, it has to be considered a technological dead end.

Another noncompression engine that achieved a measure of success was invented in Germany by Nicolaus Otto (1832–1891). This engine sold in significant numbers and brought a good income to Otto and his partner Eugen Langen (1833–1895), but in an effort to make a more efficient engine, Otto came up with the four-stroke cycle in 1876. In this type of engine, the downward movement of a piston creates a partial vacuum that sucks into the combustion chamber a mixture of air and vaporized fuel that is admitted by the opening of a valve. The mixture is then compressed by the upward movement of the piston. The actual work is done during the third stroke of the cycle, when the ignition of the fuel-air mixture pushes the piston down and turns the crankshaft to which it is connected. In the fourth and final stroke of the cycle, the piston moves up again, ejecting the exhaust gases from the combustion chamber through a port that has been opened by another valve.

As the preceding description indicates, what came to be called the "Otto cycle" is hardly a model of engineering elegance; four rapidly accelerating and decelerating up-and-down motions of a piston are required to produce a single power stroke and the rotary motion that eventually make the wheels go around. Its inventor thought it a stopgap measure and that he would eventually come up with something better, but there was no denying that the engines were a technical and commercial success; within ten years Otto's firm had sold 45,000 of them as stationary power plants for a variety of industrial uses.

## THE FIRST AUTOMOBILES

Two of Otto's employees, Gottlieb Daimler (1834–1900) and Wilhelm Maybach (1847–1929), thought that the four-stroke engine might be used to power a road-going vehicle. In 1885 they installed a single-cylinder 0.5-hp engine in the wooden frame of a two-wheeled vehicle. It is often described as the first motorcycle, but it was merely intended to demonstrate the practicality of the relatively small and lightweight engine they had created, and was not intended to be ridden any significant distance. During the following year Daimler and Maybach used the same type of engine to power a wagon, the first four-wheeled vehicle powered by an internal combustion engine.

This vehicle, it can be argued, was the first automobile, but a year earlier, while Daimler and Maybach were building their "motorcycle," in another part of Germany, Carl Benz (1844–1929) was constructing a three-wheeled car that employed an engine he designed on the principles laid down by Otto. Despite having only three wheels, it was a more technically sophisticated design in some ways. Whereas Daimler and Maybach based their car on a conventional horse-drawn wagon, Benz's employed a specially built tube frame. Its engine, however, was not as advanced, producing its 0.75 hp at 250 rpm while Daimler and Maybach's ran at a giddy 750 rpm.

It wasn't long before the practicality of the three-wheeler was demonstrated. Unbeknownst to Carl Benz, his wife Bertha and their two teenaged sons decided to take it to Pforzheim, a distance of about 60 miles (100 km). Although some hills required them to get out and push, they arrived at their destination, and a few days later they made the return journey home. Some mechanical problems had been encountered en route, but none proved insurmountable. The age of the road trip had begun.

Although Germany had been the birthplace of the automobile, France was the scene of its most important early development. Although some Daimler and Maybach cars were assembled under license in France, credit for constructing the first car designed in France goes to Peugeot, a firm that had long produced sheet steel and various metal products, including bicycles. Its first car, constructed in 1890, used a V-twin (two cylinders arranged in the shape of a "V") Daimler engine that had been built under license by another French firm, Panhard et Levassor. In the same year the latter firm followed with a car of their own design. Of the two, the Peugeot was more advanced, in that it used independently pivoting wheels, unlike the Panhard, which was steered by front wheels attached to an axle that pivoted in the middle.

At this point all automobiles mounted their engines between the front and rear wheels. In 1891 Panhard et Levassor introduced a car based on what came to be called the Système Panhard: a front-mounted engine that turned the rear wheels through a driveshaft that ran under the car's floor. This arrangement set the pattern for many decades because it allowed the use of larger engines and provided some measure of isolation from all the commotion caused by early power plants.

## A SLOW START IN THE UNITED STATES

While German and French pioneers were making rapid strides in automobile development, the United States remained stuck in the horse-and-buggy era. This is a bit curious, since by this time the United States had become a major industrial power, demonstrating a particular knack for making large numbers of products like typewriters, watches, and firearms at low cost. It might be thought, therefore, that it would be among the leaders in the design and production of automobiles. Such, however, was not the case, at least not at first. The first automobile powered by an internal combustion engine was built in 1893 in Springfield, Massachusetts, by Charles (1861–1938) and Frank Duryea (1869–1967), who had been inspired by a *Scientific American* article about the Benz car. By 1898 they had produced thirteen cars for sale to the general public; in contrast, Benz sold 600 cars in 1899 alone (Bird 1967, 56). Other early manufacturers did somewhat better. At the turn of the century, the leading American firm was Locomobile, which produced 5,000 automobiles from 1899 to 1902, but their cars were steam powered. The first U.S. manufacturer to produce significant numbers of cars powered by internal combustion engines was Oldsmobile, which can be credited with 425 sales in 1901 and more than 5,500 three years later. The "Merry Oldsmobile" immortalized in a popular song was a simple affair, and technologically retrograde in comparison with what the French were making at the time. Mounted under the floor and between the

The "Merry Oldsmobile." In addition to being one of the first mass-produced automobiles, the Oldsmobile was the subject of a popular song. Courtesy of the Lester S. Levy Collection of Sheet Music, Special Collections, The Sheridan Libraries of The Johns Hopkins University.

front and rear wheels of its buggy-like body was a single-cylinder engine that put out 7 hp, giving the car a top speed of about 20 mph. Given its primitive brakes and tiller steering, this was surely fast enough.

While Oldsmobile was enjoying a measure of commercial success, other would-be manufacturers were struggling to design cars that would meet minimally acceptable standards of performance and reliability. The apparent practicality of automobile travel was demonstrated when several early manufacturers engaged in long-distance journeys. What has been described as the first American automobile race took place in 1895, when five cars lined up on a snowy November day in Chicago. Two of them, a Benz and a Duryea, made it to their destination 55 miles away. Although the winning Duryea averaged less than 8 mph, a newspaper report pointed out that the race had been run "through deep snow, and along ruts that would have tried horses to their utmost" (quoted in Flink 1970, 23). In 1897 Alexander Winton drove one of his cars from Cleveland to New York, a distance of

800 miles that was covered in 79 hours of driving time. In 1901 an Oldsmobile was driven from Detroit to New York, where it was displayed at an automobile show held in that city. This trip took seven days, as did the journey of a Haynes-Apperson from Kokomo, Indiana, to New York in the same year. Another Winton achieved a significant first in 1903 when H. Nelson Jackson and Sewall Crocker drove it coast-to-coast from San Francisco to New York, a trip that took 63 days. The same year saw a Packard travel the same route in 53 days, while an Oldsmobile did it in 65 days (Rae 1984, 24), all of this occurring at a time when the United States had fewer than 200 miles of paved roads.

## GASOLINE, ELECTRIC, OR STEAM?

By this point the gasoline-fueled internal combustion engine was generally accepted as the best means of powering an automobile. Why this was so requires some explanation. When cars first began to be sold during the waning years of the nineteenth century, it was not at all obvious that the internal combustion engine would emerge as the preferred source of power. In fact, of the 4,192 cars produced in the United States in 1900, 1,681 were steamers, 1,575 were electrics, and only 936 used internal combustion engines (Flink 1988, 234). The two rivals of the gasoline engine certainly had their virtues. Steam technology was established and well proven. A steam engine delivered peak torque at low rotational speeds, eliminating the need for a geared transmission. Steam engines were reasonably quiet, and they vibrated less than early internal combustion engines. And they were fast, as evidenced by the speed of 127.7 mph (204.3 kph) attained by a streamlined Stanley at Ormond Beach, Florida, in 1906. There even existed the legend that the Stanley brothers would give a new car to anyone who could hold open the throttle of one of their steamers for more than a minute. In contrast, high speed was not a virtue of electric cars. But electrics were agreeably quiet, were easy to drive because there were no gears to shift, did not vibrate, emitted no unpleasant odors, and did not have to be cranked to life.

Offsetting these virtues were a number of drawbacks. Steam cars did not require cranking, but it took several minutes of heating to turn water into steam. The "flash boiler" invented in France by Leon Serpollet reduced the time to 30 seconds, but operators still had to exercise considerable advance planning and continuous oversight to prevent their vehicles from running out of steam. Steamers also had a prodigious appetite for water; in 1900 a Stanley-designed Locomobile performed the impressive task of driving from one corner of Britain to another, but it consumed five tons of water in the process (Bird 1967, 120). Condensers, which were rarely fitted to steam cars, allowed water to be reused, but they added weight and complexity, and were prone to scaling because the steam they condensed carried lubricating oil for the engine. Finally, because even a superheated steam engine cannot operate at the temperatures attained by internal combustion engines, thermal efficiency suffered, as was reflected in the high fuel consumption typical of steamers.

Electric cars were well suited for around-town jaunts, but their overall performance was hampered by the inherent limitations of using batteries as a source of power. Batteries were heavy and needed frequent recharging, a difficult task given the underdeveloped technology for converting commonly used alternating current into direct current. In 1899 a specially built electric car was the first automobile to exceed 60 mph, but electric passenger cars topped out at about 20 mph and had a range of about 60 miles under good conditions.

Cars powered by internal combustion engines had their faults, too. Early cars with their engines residing under the seat put the driver and passengers in close proximity to a great deal of mechanical uproar, and as one early advocate of electric cars noted, "You cannot get people to sit over an explosion" (quoted in Rae 1984, 15–16). Many purchasers of cars powered by internal combustion engines disagreed, but they could not have been pleased by the vibration, noise, and fumes that were inescapable companions of the combustion of gasoline. And most disturbing of all was the prospect of getting the thing started by swinging a crank, a task that struck fear into the hearts of many early motorists. And once he or she was underway, the motorist could only hope that their journey would not be terminated by a leaking carburetor, a burned valve, or an electrical coil that mysteriously stopped functioning.

Yet with all these seemingly crippling drawbacks, by the first decade of the twentieth century the internal combustion engine had triumphed. Electric cars retained a small niche for in-town transportation, and a few hundred steamers were sold each year, but these were strictly sideshows. The reason for the dominance of the internal combustion engine has been the subject of considerable debate. Some of the explanations given for the demise of steam cars, like the claim that they were sabotaged by the oil industry, are ill-informed; after all, the boilers of steam cars were heated by the combustion of gasoline—lots of it. Others have claimed that with more development, steam-powered cars would have demonstrated their inherent superiority, but this can be filed under the category of "what might have been." In regard to electric cars, some have laid the blame at the feet of the automobile industry for failing to put their engineering and marketing weight behind them, but even electrical power companies, which might have seen electric car owners as important customers, showed little interest in promoting the cars.

Putting things more positively, the reasons for the dominance of the internal combustion engine and the eclipse of its rivals can best be understood by placing the automobile in its social, cultural, and even psychological context. Early automobiles were never intended to serve as everyday transportation devices. They were all sports cars in the sense that speed, adventure, and the appeal of technological novelty were among their key attractions. This requirement effectively eliminated the electric car, which offered little in the way of excitement. Making matters worse, electrics were identified as "ladies' cars," which made them inherently unacceptable to the majority of drivers, who happened to be male. Then as now, it was widely believed that "you can sell a man's car to a woman, but you cannot sell a woman's car to a man." Steam cars offered a high standard of performance, but their operation required a great deal of driver involvement, with gauges to watch, valves to be adjusted, and in general the maintenance of a high level of vigilance. This may have appealed to drivers who went in for this sort of thing, but it was too much for the average driver. All in all, the gasoline-fueled internal combustion engine offered the best compromise. It combined high-performance potential with reasonable practicality, and the rapid strides made in improving efficiency, quelling vibration, and simplifying operation made it even more acceptable for average motorists.

## THE TECHNOLOGICAL DEVELOPMENT OF THE AUTOMOBILE

The triumph of the internal combustion engine as the automobile's source of power was paralleled by the standardization of its other major components. Considerable variation occurred from model to model, and a great deal of improvement took place over the years, but the basic layout of the automobile that was set down during the first decade of the twentieth century endured for many decades thereafter. As noted above, the great majority of automobiles were designed along the lines of the Système Panhard—a front engine supplying power to the rear wheels through a clutch, a sliding-gear transmission, and a differential gear. But a great deal of development took place within this basic plan, so that by 1905 automobiles were far superior to what had been found on the roads just five years earlier.

The sporting nature of early automobiles was reflected in their rudimentary bodywork. Skimpy fenders retarded the spray of dust, water, and mud, but the weak engines of the day did not allow the fitment of anything more substantial. Traditional carriage makers were sometimes called upon to supply bodies, but in most cases these were too heavy, and some creations were nothing short of ridiculous, as when coachbuilders tried to emulate the elegant horse-driven conveyances of times past. Even the best designs had a bits-and-pieces look about them; engineers and technicians, and not stylists, were responsible for the look of a car, and it wasn't until the 1920s that cars began to be consciously designed as integrated entities. Throughout the first decade of the twentieth century the vast majority of cars still had open bodies, and even windshields were not universal fittings at first. Consequently, neck-to-toe "dusters," blankets, fur coats, caps, gauntlets, hoods, and goggles were common apparel for motorists and their passengers. The few closed cars of the time were for the most part intended for formal occasions, and a large measure of their ungainliness can be attributed to the high roofline needed to accommodate gentlemen in top hats and women wearing elaborately decorated hats as they drove to the opera or some other gala event.

The construction of auto bodies still owed its basic techniques to the carriage industry. Supportive frameworks were generally wooden, over which metal panels were fastened. Painting was a laborious process, entailing many coats of sealer, primer, and body color, all of which had to be sanded smooth before the next coat could be applied. Fittings such as lamps, control levers, and horns were usually brass, and maintaining their sheen required continuous attention. Headlights required a great deal of maintenance beyond keeping their housing shiny; the most effective ones used acetylene that was generated by dripping water over calcium carbide crystals. This left behind a messy residue that had to be cleaned out at regular intervals. Electric lighting was occasionally employed, but the lack of suitable electrical generators limited their adoption until about 1910.

Although they were far more maintenance intensive than today, by the middle of the twentieth century's first decade, cars had reached a point where motorists had a decent chance of reaching their destinations and returning without suffering any mechanical trauma. If problems detained them it was likely the fault of the tires, which were prone to punctures and other failures. A service life of about 1,000 miles was all that could be expected.

The very first cars rode on solid-rubber tires; the pneumatic tire had been invented back in 1845 by Robert Thompson, but they were forgotten until the bicycle created a vast market for tires that would absorb road shocks and dampen vibration. The pneumatic tire was thus reinvented in 1888 by John B. Dunlop (1840–1921), a Scottish veterinarian living in Northern Ireland. The first use of pneumatic tires for an automobile is attributed to the Michelin brothers, who used them for a Peugeot that participated in the 1895 Paris-Bordeaux-Paris race. Numerous punctures put them last across the finish line, but they had been able to finish the race, something that thirteen of the starting field of twenty-two had failed to do. Tires gradually became more durable, which was a good thing because repairing a flat tire was not an easy task. First, the tire had to be wrestled off the wheel because the latter was permanently attached to the car. After a puncture was mended or a tube replaced, the tire had to be mounted back on the wheel, and the operator had to use a hand pump to inflate the tire to pressures of up to 100 pounds per square inch (PSI). It was not until well into the first decade of the twentieth century that demountable rims and then detachable wheels removed some of the terrors induced by pneumatic tires.

The first automobiles were powered by one- and two-cylinder engines. Four-cylinder engines appeared in 1896, and within a few years they had become the standard engine configuration. By a process of extrapolation a six-cylinder engine could be expected to run more smoothly because at a given displacement each piston could be smaller and lighter, but in fact torsional vibration caused by a longer crankshaft negated this advantage. Within a few years the vibration damper invented in England by H. F. Lanchester (1868–1946) cured the problem, but six-cylinder engines remained largely confined to the most expensive automobiles.

Almost every internal combustion engine used poppet valves for the intake of a fresh charge of air and fuel, and the exhaust of a spent one. Exhaust valves were operated by a camshaft driven by the crankshaft, but until the first few years of the twentieth century most engines employed "automatic" or "atmospheric" intake valves that were pulled open by the suction created by the downward movement of the piston, and then closed by a light spring. This arrangement had the virtue of simplicity, but the early closure of the valve prematurely cut off the flow of the fuel-air mixture, with negative consequences for power and efficiency. Getting the proper tension on the spring was a matter of compromise; if it was too tight the valve opening was delayed, while if the spring was too weak the valve could bounce and eventually break. Mechanically operated valves thus became the norm; they were generally located along the side of an engine, but a few designs made use of overhead valves that opened into the combustion chamber from above. This was a more complicated arrangement that necessitated pushrods and rocker arms, but it made for better combustion chamber design and consequently more efficient operation.

Early automobile engines often reflected the fact that their designers got their start working with stationary engines. Most early automotive engines were designed to operate over a narrow range of rotational speeds, and many lacked any means of control other than overriding a governor that limited the engine's speed. Atmospheric intake valves were partly responsible for the limited flexibility of these engines, but the situation was exacerbated by the absence of effective carburetion. The first engines used an apparatus that simply directed air over the surface of the fuel in order to mix the two. This hit-or-miss affair was rendered obsolete by Wilhelm Maybach's 1893 invention of the ancestor of all carburetors, which used a float to control the influx of fuel, which was then mixed with air as it sprayed out of a jet. This was an improvement, but a great deal of development work lay ahead before carburetors provided properly vaporized fuel (instead of emitting a stream of droplets) and a precise metering of a correct air-fuel ratio.

Many early cars used a heated tube to ignite the air-fuel mixture. In hot-tube ignition, one end of a platinum tube was mounted in the cylinder head while the other end was kept hot by a burner. When a fresh charge of air and fuel entered the combustion chamber, some of it was forced into the tube, where it ignited; this in turn caused the ignition of the air-fuel mixture in the combustion chamber. Hot-tube ignition had the virtue of simplicity, but in addition to posing a fire hazard it did not allow the timing of ignition to be varied. Again, this was not a serious drawback when engines ran at something close to constant speed, but by 1900 the advantages of more flexible engines had rendered hot-tube ignition obsolete. In its place, early automobiles were served by two basic types of electrical ignition. The first was a battery-and-coil arrangement that had the advantage of delivering a strong spark, but it required an effective and fully charged battery. In a modern car the battery is recharged automatically by an onboard generator or alternator, but difficulties in regulating the output of voltage prevented their use until the second decade of the twentieth century. In order to obviate the need for batteries, many early cars used a magneto, which induces an electric current by rotating magnets past stationary coils. This had the virtues of simplicity and reliability, but it delivered a weak spark at low engine revolutions, which could make starting difficult.

As with many features of early automobiles, suspensions initially were based on those employed by horse-drawn vehicles. Leaf springs, usually of the semi- and quarter-elliptic variety, were most commonly employed.

These were usually arranged in a fore-and-aft fashion, although some cars used a transverse semielliptic at the rear. Coil springs were less popular, in part because multileaf elliptical springs provided a measure of damping, a useful quality at a time when spring dampers (misleadingly called "shock absorbers" today) were a rarity.

Brakes were barely adequate, especially when the speed capabilities of automobiles rapidly increased. Brake linings were leather or fabric, and as can be imagined, their life expectancy was short. Woven asbestos brake linings appeared around 1905, but a number of years passed before they became common. Front-wheel brakes were rarely employed, but many cars had a dual-braking system, typically with a pedal-operated contracting band that gripped some part of the transmission or the driveshaft, and a hand-operated "emergency" brake that actuated shoes that pressed on rear drums. Since deceleration transfers a car's weight forward, the lack of brakes on the front wheels seriously limited stopping power, but automobile engineers were not confident that brakes could be safely applied to the wheels that steered the car.

Most engines used a liquid cooling system. At first, the water was cooled by circulating it through a set of large pipes festooned with fins. A better system appeared on the 1901 Mercedes, which employed what came to be called a "honeycomb" radiator. This was based on a vertical tank that contained hundreds of small air tubes that cooled the water that surrounded them. This became the standard arrangement, and the radiator, often brass-plated and topped with an ornamental "mascot" perched on the filler cap, became the most prominent feature of a car's front.

## THE FIRST MOTOR RACES

As an old cliché would have it, "Racing improves the breed." Whether or not this is still true in regard to automobiles is debatable, but it is evident that racing made significant contributions to early automotive progress. The first organized contest, more a reliability run than an out-and-out race, occurred in 1894; it began in Paris and ended in Rouen, a distance of 78 miles (126 km). The next year a longer course was traveled, from Paris to Bordeaux and back. It was won by a Panhard et Levassor that covered 732 miles (1,178 km) at an average speed of nearly 15 mph. Town-to-town races, some of them crossing international borders, became all the rage until they met their tragic culmination in the 1903 Paris-Madrid race. By this time racing cars were capable of top speeds of more than 75 mph, far in excess of the capabilities of the brakes and suspensions of the time. The result was a plethora of lethal accidents that caused the race to be terminated at Bordeaux. The exact death toll has never been determined, but it was probably in the neighborhood of fifteen, six of them drivers. The race was won by a Mors driven by Fernand Gabriel, who averaged an astonishing 65.3 mph (105 kph) for 342 miles (550 km). High speeds, coupled with the impossibility of keeping large crowds under control, put an end to these city-to-city races. In their place, race organizers promoted contests that took place on closed circuits that employed portions of public roads or were staged on specially built tracks like England's renowned Brooklands raceway.

Early race cars were closely related to road cars, but specialized racing machines began to appear during the early 1900s. Exemplifying the old racing maxim that "there's no substitute for cubic inches," some of these vehicles employed gigantic engines, the biggest of which was the 28-liter (1,700 cubic inches) monster that propelled a Fiat racer to speeds exceeding 150 mph at the Brooklands track. Engines of this sort had more relevance to the emerging aircraft industry than they did to ordinary road vehicles.

The early interest in racing, both on the part of drivers and spectators, shows almost from the outset that the automobile was viewed as more than a utilitarian device for transporting people between two points. Beginning with Daimler and Benz, many automotive pioneers were motivated by "technological enthusiasm," the desire to create a technological novelty for the sheer pleasure of making it work. Other early automobile manufacturers got into the business because they saw the opportunity to profit from a rapidly growing market, but in many cases both motivations were closely intertwined. In a number of ways, the history of the early automobile industry resembled the personal computer boom of the 1980s and the Internet mania of the 1990s. Entrepreneurs were excited by the prospect of making a lot of money, but at the same time the product itself conveyed an excitement that was not present in more established products, resulting in the entrance of many firms with little hope of long-term survival.

Whatever the source of their motivation, early entrepreneurs faced few obstacles to getting into the automobile business. In comparison with what came later, cars were simple—a chassis, a drivetrain, and some sort of bodywork. No safety or emissions standards had to be met, and customers were undemanding. A car was deemed acceptable if it could hit 30 or 40 mph, climb most hills—however slowly—and offer its operator a decent chance of returning from most journeys.

Automobile manufacture also did not require vast amounts of capital, intellectual or financial. The know-how necessary to design an adequate car was often developed through experience and practice, although university-trained engineers were well represented in the early industry. Very few of a car's components had to be designed and constructed by the manufacturer. For example, the first cars made by Panhard et Levassor used Daimler engines, and a few years later the former firm supplied the engines used in the first Peugeots. Even when a component was designed by the manufacturer, it might be built by another firm. To take one example out of many, early Oldsmobiles had engines made by one outside firm and transmissions by another. Accessories such as lights and electrical gear were almost always made by outside suppliers.

When automobile companies made their own parts, they often benefited from prior manufacturing experience. As was noted above, many automobile manufacturers started with bicycles. One interesting example is Pierce, which began in Buffalo, New York, as a manufacturer of bird cages. This led to the production of wire wheels for bicycles, then complete bicycles, and finally to automobiles that eventually gained an outstanding reputation for luxury and high quality. Other manufacturers began by making such varied products as bathtubs (Buick), washing machines (Peerless), and sheep-shearing equipment (Woolsley). Perhaps the oddest route into automobile manufacture was taken by the Smith Automobile Company of Topeka, Kansas, which began with the manufacture of hernia trusses.

Whatever their point of entry, early automobile manufacturers could get into the business with only modest financial backing. Components obtained from outside suppliers could be purchased on credit, but finished cars were sold for cash to dealers and distributors, who often had to put down a substantial deposit in order to place an order for cars yet unbuilt. Thus the Ford Motor Company, which rapidly became a major force in the industry following its formation in 1903, got started with an investment of only $28,000 (Flink 1970, 295). Low capital requirements allowed literally hundreds of firms to get into automobile manufacture during the first decade of the twentieth century, but fewer than half survived, and most of them did not make it through the next decade.

The key role played by component suppliers helps to explain the geographic concentration of the automobile industry; once a "critical mass" had been reached, manufacturers could draw on the products and services of a wide range of local component suppliers. Although cars were built in many parts of the United States, the industry tended to cluster in a few areas, notably Cleveland, Buffalo, Indianapolis, and Detroit, with the latter eventually becoming the center of the industry. Why Detroit became the preeminent "motor city" has been a matter of some debate, but most cred-ible is John Rae's assertion that

> the real explanation comes down to people, to the fortuitous circumstance that a remarkable group of automobile entrepreneurs appeared simultane-ously in the Detroit area. . . . Together they made Detroit the principal cen-ter of motor vehicle production in the nation and eventually in the world, and once the concentration was established it grew on itself. Detroit came to exercise a magnetic attraction on everyone, individual or organization, who wished to be involved in the manufacturing of motor vehicles. (1984, 29–30)

## THE AUTOMOBILE'S CLIENTELE

Many early manufacturers failed, some after a year or two, others after the passage of decades. The reasons for their failure were varied, but the lack of potential customers was not one of them. Automobile ownership grew rapidly during the first decade of the twentieth century, and in 1904 an important threshold was crossed when the American manufacturers pro-duced more than 22,000 cars, allowing the United States to surpass France as the world's largest producer of automobiles. Three years later, American automobile production was greater than all of Europe's combined (Mc-Shane 1994, 108). American numerical superiority, which endured for many decades to come, was driven by the forces of both supply and de-mand. On the supply side, American manufacturers could draw on a wealth of experience in industrial processes, and the United States' indige-nous labor force was supplemented by large numbers of recent immi-grants, a fair number of whom possessed crucial technical skills. Many American industries had already led the world in the production of inex-pensive, mass-produced goods, and during the early twentieth century the automobile industry would demonstrate an unparalleled ability to manu-facture large quantities of the most complicated consumer product made up to that point.

At first, the consumer base for automobiles was limited. In both Europe and the United States, automobile ownership was initially an urban phe-nomenon due to the availability of decent roads and the presence of a population with money to spend. The latter was particularly important during the opening years of the twentieth century when cars were quite expensive in comparison with established commodities. At a time when

American factory workers earned about $2 a day, or perhaps $600 a year, as unpretentious a car as the one-cylinder Oldsmobile Runabout would have absorbed a worker's entire annual income. Cars offering more luxury and performance such as a four-seat Winton carried a price tag of $2,500, and for the truly affluent an imported Mercedes could be purchased for a cool $12,750. Consequently, buyers of early automobiles tended to come from the ranks of the upper and upper middle classes. Engineers and businessmen were well represented among the ranks of early car owners, as were physicians; in that long-gone era when doctors made house calls, an automobile allowed more patients to be seen in a day than was possible with a coach and horse.

With car ownership beyond the financial reach of ordinary workers, many of them saw the automobile as the embodiment of an arrogant and self-centered moneyed class. As Woodrow Wilson, then serving as the president of Princeton University, noted with disdain in 1906, "Automobilists

Chicago Street. Early automobiles had to share crowded city streets with pedestrians, streetcars, and a variety of other vehicles. Courtesy of the Library of Congress.

are a picture of arrogance and wealth, with all its independence and carelessness . . . nothing has spread socialistic feeling in this country more than the automobile" (quoted in Brandon 2002, 55). Cars met with particular hostility in densely settled cities, where people living in crowded tenements used the streets for games, socializing, buying and selling, and other activities that had nothing to do with transportation. In this environment, the appearance of an automobile represented a threat to an accustomed way of life, while at the same time posed a very real safety hazard.

Automobiles also raised the ire of countryfolk who complained about reckless drivers, the dust kicked up by cars, and the proclivity of motorists to trespass on private property, steal fruit, and strew trash. In extreme cases, rural residents responded to the invasion of their territory by digging up the roads or stretching barbed wire across a highway.

Hostility toward the automobile quickly subsided as car ownership expanded at a rapid pace in the United States. Once the automobile had demonstrated its reliability and practicability, and prices began to come down, the industry was assured of a large market because the United States had the world's highest per capita income, which, unlike today, was more evenly distributed than was the case in Europe. The result was a rapid rise in the ownership of automobiles, as Table 1.1 indicates.

This was only a hint of what was to come. As will be seen in the next chapter, a revolution in the manufacture of cars lay just over the horizon, giving rise to an explosive growth of automobile ownership.

**Table 1.1**
**Early Twentieth-Century**
**Automobile Sales in the**
**United States**

| | |
|---|---|
| 1900 | 4,192 |
| 1901 | 7,000 |
| 1902 | 9,000 |
| 1903 | 11,235 |
| 1904 | 22,130 |
| 1905 | 24,250 |
| 1906 | 33,200 |
| 1907 | 43,000 |
| 1908 | 63,500 |

## GOVERNING THE AUTOMOBILE

The rapid diffusion of automobile ownership in the United States was not matched by government involvement in shaping the social, economic, and geographic landscape that was being created by the automobile. In contrast, European governments set down many regulations governing the use of automobiles at an early date. The most extreme dated back to 1865 when Britain enacted the Locomotives on Highways Act, commonly known as the Red Flag Act because it required a self-propelled road vehicle to be preceded by an individual waving a red flag during the day and a red lantern at night. The act also restricted these vehicles to 4 mph (7 kph) in the countryside and 2 mph (3 kph) in towns. A revised act, passed in 1896, eliminated the flag and lantern, and elevated the speed limit to 14 mph (20 kph), still unrealistically low. Further hindering automotive development was a British law that forbade the use of highly volatile fuels (e.g., gasoline), condemning internal combustion engines to the use of kerosene until the law was repealed in 1900. Despite Britain's industrial preeminence, automotive development languished under these conditions, and ceded to France and the United States leadership in the production and marketing of automobiles during the early twentieth century.

Unlike the situation in many parts of Europe, the decentralized nature of American government precluded a unified approach to regulating the automobile. Many states did not even require the registration of cars. New York was the first state to mandate registration, doing so in 1901, but it wasn't until 1915 that all states required that cars be registered, and even then many of them were satisfied with only an initial registration. It was not until 1921 that all states mandated annual registration. The licensing of drivers, well established in Europe, also came relatively late to the United States. As the first decade of the twentieth century came to a close, only twelve states and the District of Columbia made driver's licenses mandatory. And even in these instances, the applicant simply had to provide some basic information; no written test was required, much less a driving test. Speed limits, when they existed, were set by towns and municipalities, although it was out in the countryside where high speeds were most likely to be attained. By 1906 most states had gotten around to setting speed limits for open highways. Of these limits, the highest was 24 or 25 mph, found in only five states. Alabama had the lowest, allowing its motorists a top speed of only 8 mph. As might be expected, many local governments were quick to seize on the revenue potential of issuing speeding tickets, and the exploitation of "speed traps" became a common motorists' complaint.

From a vantage point of 1905 or thereabouts, an observer could look back on the twenty years that had passed since the invention of the automobile and come away impressed by the substantial progress that had taken place. The automobile was no longer an experiment conducted by a few technological enthusiasts, but a practical and reasonably reliable means of transportation that was winning large numbers of new adherents every day. Racing cars had reached fantastic speeds, while ordinary passenger cars were going on journeys that would have been unimaginable a few years earlier. An industry that scarcely existed a few years earlier was turning out models of every description, and each year saw substantial improvements in performance, comfort, and durability. The automobile had reached a certain level of maturity, but its most expansive years lay ahead.

2

# The Automobile's Adolescence, 1905–1914

Although the invention and early development of the automobile took place in Europe, by the first decade of the twentieth century the United States had become the leading automotive nation. American cars may not have been at the forefront technologically, but no nation surpassed the United States in the invention and use of advanced productive technologies. Yet while cars poured out of American factories, motorists continued to be frustrated by appallingly poor roads, a shortcoming that was to remain for many years to come. Even so, the limitations of early automobiles and the roads that they traveled on did not deter rapidly growing numbers of people from becoming part of the motoring public. A fair proportion of these new motorists were women, and their presence behind the wheel was only one of many social and cultural changes that were stimulated by large-scale automobile ownership.

## THE UNITED STATES TAKES THE LEAD

In 1904 American automobile manufacturers produced 22,130 vehicles, allowing the United States to pass France as the world's largest manufacturer of automobiles. In the years that immediately followed, the production and ownership of cars in the United States expanded at a breathtaking rate; in all but two years during the span from 1905 to 1914, the production of automobiles in the United States increased by 20 percent or more (Bardou et al. 1982, 51). Much of this increase can be attributed to America's prior leadership in the production of massive quantities of industrial goods. At the same time, social and economic conditions in the United States created optimal conditions for widespread automobile ownership. Most importantly, at the beginning of the twentieth century incomes in the United States were considerably higher and more evenly distributed than they were in Europe. In 1914 per capita income in the United States was $335, considerably above the figure of $243 for Great Britain, $185 for France, and $146 for Germany (Bardou et al. 1982, 47). Automobiles have always made significant claims on the financial resources of their owners, and relatively high average incomes are a prerequisite for widespread automobile ownership.

While purchasing power was relatively high, the composition and culture of the American industrial work force were making it easier to lower the price of cars through the use of mass-production technologies. While American manufacturing techniques often required little in the way of workers' skills, European industrial production heavily relied on the efforts of high-priced skilled workers. To take one example, of the 1,700 production workers employed by Daimler in 1909, 1,175 were skilled, 200 were semiskilled, and 325 were unskilled (Bardou et al. 1982, 63). Altogether, they produced fewer than 1,000 cars annually. The factory was the domain of skilled craftsmen, and they guarded against anything that threatened their autonomy and privileges. Consequently, the employment of unskilled workers and even modern machine tools often met with resistance. Reliance on skilled workers was appropriate for the manufacture of sophisticated, high-priced cars favored by European manufacturers, but it impeded the development of large-scale, standardized production.

The combination of the lack of a mass market and traditional production methods meant that European cars were largely designed and manufactured with the affluent and the upper middle classes in mind. In contrast, the story of automobile production in the United States during this period centers on efforts to expand car ownership by driving down production costs and making automobile ownership possible for millions of people. The embodiment of this strategy was the Ford Motor Company.

## THE ASSEMBLY LINE

According to a story that may be apocryphal, late in his life Henry Ford (1863–1947) was being interviewed by a young reporter. At one point the reporter noted that the views of the cantankerous industrial giant were perhaps out of step with the modern world. At this point, the old man cut him short by asserting, "Young man, I invented the modern world."

There is considerable truth in this seemingly outrageous statement. Perhaps more than any other twentieth-century figure, Henry Ford was responsible for the interlinked economic, technological, social, and cultural changes that produced what we think of as modern society. Not only did he take the lead in producing automobiles that gave mobility to tens of millions of people in the United States and many other parts of the world, but the industrial practices that he and his associates developed also set the pattern for the mass production of goods and its necessary complement, a culture of mass consumption. Ford's accomplishments were well recognized during his lifetime, and adulation for him reached such a level that his daily mail included hundreds of letters seeking his advice on every imaginable topic. At one point he was seriously considered as a possible presidential candidate. More ominously, Ford's admirers also included two of the greatest mass murderers of the twentieth century, Josef Stalin and Adolf Hitler.

Why was Ford such a significant, even revolutionary, figure? Consider the year 1903, the year that the Ford Motor Company was founded. In that year, 11,235 automobiles were sold in the United States. Twenty years later, the number had risen to 3,624,717, a more than 300-fold increase (American Automobile Manufacturers Association 1962, 104). In 1923, a year that Ford produced more than 1.8 million cars, half of the automobiles on the road were Model T Fords, the legendary "flivver" or "Tin Lizzie." To achieve this degree of market domination, Ford's company employed a strategy of relentlessly driving down production costs and cutting prices. When it came out in late 1908, a Model T touring car retailed for $950. By 1915 its price was down to $550, and in 1924 one could be bought for only $290. By this time the nation's highways were awash with Model Ts being driven by people who in the recent past had relied on bicycles, horse-driven buggies, and crowded streetcars for their transportation needs.

Henry Ford's ascent to industrial domination began in rural Michigan, where he developed a thoroughgoing antipathy to traditional, nonmechanized farming. He found mechanical things far more appealing than farm chores, and at the age of sixteen he left home to work as a machinist, eventually rising to the position of chief engineer at the Edison Illuminating Company's Detroit facility. His first automobile was a home-built job that he put together in 1896. Dubbed the Quadracycle, according to a frequently repeated story he had to knock down a portion of a wall in order to get it out to the road. After being involved with two failed attempts to profitably manufacture automobiles, Ford and his partners founded the Ford Motor Company in 1903.

At the time of the company's formation, automobiles were either expensive devices purchased by the well-to-do, or cheap vehicles with severely limited performance and dubious reliability. Ford's partners intended to pursue a strategy of making and selling high-priced cars, but Ford was more interested in catering to a broader market. Cars would be sold at a low price, but the large volume of sales would make for substantial profits.

That at least was the theory, but putting it into action required epochal changes in the way automobiles were produced; cars had to be produced like much simpler objects by using the techniques of mass production. Although the term "mass production" was coined by Ford's ghostwriter in a 1926 entry in the *Encyclopaedia Britannica*, the idea and practice of large-volume manufacturing had been around for a long time, and its basic principles were well understood. First, mass production required a standardized product, one that could be made in large numbers with no costly and time-consuming variations. Second, mass-produced items were assemblies of interchangeable parts so that the production process was not slowed down by the need to modify or excessively manipulate the components that were

A Model T poses with Henry Ford's first automobile, which he dubbed the Quadracycle. Courtesy of the Library of Congress.

being assembled into the final product. Third, the manufacturing process made abundant use of specialized tools that were designed to do specific tasks as rapidly and effectively as possible. Fourth, the skills required to make the product resided in these tools, not the workers who used them; this meant that the workforce could largely consist of low-skilled, and hence low-wage, labor.

None of these principles originated with Ford; they had been used increasingly during the late nineteenth and early twentieth centuries for the manufacture of everything from shoes to clocks to canned food. Ford was not even the first to manufacture automobiles on a large scale. The French firm De Dion Bouton manufactured 1,500 cars over a one-year period that extended from 1900 to 1901, while Locomobile in the United States built about 5,000 steam-powered cars between 1899 and 1903. In 1901 Ransom E. Olds produced 425 curved-dash Oldsmobiles, and he sold more than 5,500 of them in 1904, more than any other manufacturer had achieved up to this point (Flink 1988, 31–32). Nor was Ford a pioneer in the use of interchangeable parts; one dramatic illustration of the use of interchangeable parts in automobile manufacture occurred in 1908 when three Cadillacs were taken to the Brooklands speedway in England, where they were completely disassembled. The cars were reassembled from the scrambled parts, and filled with gas, oil, and coolant. They then circulated around the track for a faultless 500-mile run.

Although significant examples of large-scale automobile production existed, none of them employed mass-production technologies to the extent found at the Ford Motor Company. Moreover, the firm took the lead in developing the industrial innovation most often associated with "Fordism," the assembly line. Prior to Ford, automobile manufacture was done in much the same way as with locomotives and other large objects: the vehicle slowly took shape as workers added the various components to it while it remained stationary. In contrast, the assembly line allowed (or forced) workers to stay in one place as they performed various tasks while the car or one of its major subassemblies came down the line. The moving line was first used for the assembly of a part of the Model T's magneto during the summer of 1913, and within a year almost every assembly operation was being performed on a moving line (Hounshell 1984, 247). Workers putting cars together on an assembly line usually had one task to perform, and very little time to do it, requiring them to repeat a simple operation all day long. As Ford explained how work was done on the line,

In the chassis assembling [line] are forty-five separate operations or stations. The first men fasten four mudguard brackets to the chassis frame; the motor arrives on the tenth operation and so on in detail. Some men do only one or two small operations, others do more. The man who places a part does not fasten it—the part may not be fully in place until several operations later. The man who puts in a bolt does not put on the nut; the man who puts on a nut does not tighten it. (Ford and Crowther 1922, quoted in Chandler 1964, 40)

The assembly line did not come completely out of the blue; precedents can be found in flour milling, brewing, and the manufacture of tin cans, as well as the use of "disassembly lines" in meat-packing plants. Moreover, most of the actual work of setting up and perfecting the line came not through Ford's own efforts, but those of his lieutenants. But there is no question that the firm led by Henry Ford was the first to demonstrate the value of the assembly line for the mass production of automobiles and other complex consumer goods.

The use of the assembly line and other elements of mass production drove down costs substantially, but they took a heavy toll on the workers who were tied to one spot, performing the same job again and again, and within the tightly circumscribed period of time dictated by the speed of the line. Working conditions were further degraded by the incessant noise that accompanied factory production. As described in Ford's official history, "the steady hum of the lathe, the incessant tapping of the hammers, the dull thud of the presses, the click-clack of the shapers, the whirr of the drills, [and] the groaning and clicking of the drilling machines and reamers" produced "a semi-hypnotic state from which the workman's mind emerges only at intervals" (Nevins and Hill 1954, 528). The consequences of the assembly line for Ford workers were poignantly described by a line operative's wife, who in a letter to Ford lamented, "The chain system that you have is a *slave driver. My God!*, Mr. Ford. My husband has come home & thrown himself down & won't eat his supper—so done out! Can't it be remedied? . . . . That $5 a day is a blessing—a bigger one then you know but *oh* they earn it" (quoted in Hounshell 1984, 259).

The $5 daily wage that her husband earned was another of Ford's innovations. In 1914 Ford announced that his workers would receive this astonishing wage (equivalent to a daily wage of about $90 today) for an eight-hour day, approximately double the prevailing rate for factory work. In this way, Ford offered to compensate for the physical and mental exhaustion engendered by assembly-line work by paying high wages. Ford also made heavy use of immigrant workers, for whom high-paid assembly-line work was preferable to serving as a farm laborer or poorly paid artisan in the Old Country. The firm also operated what it called the Sociological Department. Headed by a former minister, it had the task of helping workers to cope with their daily lives, but as we shall see, it ended up being used for far less beneficent purposes.

As Ford freely admitted, his motives in dramatically raising wages were not entirely philanthropic. The year before the $5-per-day wage was instituted, turnover among Ford employees had reached 370 percent; 71 percent of new hires departed after fewer than five days (Brandon 2002, 105). The costs necessitated by constantly hiring and training workers had eaten into profit margins to the point that Ford was able to claim that because it dramatically cut labor turnover, "The payment of five dollars a day for an eight-hour day was one of the finest cost-cutting moves we ever made" (quoted in Flink 1975, 90).

## THE MODEL T

The car that Ford workers produced embodied few technological innovations, but it was sturdy, reliable, and easy to drive by the standards of the time.

A Ford assembly line. Courtesy of the Library of Congress.

The car's designers made abundant use of vanadium steel, which gave the Model T considerable strength despite its light weight of about 1,500 pounds. Its two-speed planetary transmission obviated the need to operate a conventional gearbox and clutch, which required considerable skill on the part of the driver to avoid grinding gears. Its top speed of about 40 mph was modest, but adequate for the roads of the time and in keeping with the car's limited braking power. Its simple construction endeared it to drivers who could not enlist the services of professional mechanics. Even major repairs could be undertaken by owners who familiarized themselves with their cars while performing maintenance functions at regular intervals; some components required the ministrations of a grease gun every 500 miles. And if giving mobility to millions of people was not enough, the Model T could be adapted to other uses. Through the fitment of suitable accessories, a Model T could be used to plow a field, saw wood, or power a washing machine, although such adaptations were discouraged by Ford because they put stresses and strains on the car for which it had not been designed. The value of a Model T inhered in the mobility it conferred. It had its quirks, and Ford persisted in making it for too long, but for many years the Model T was the right car at the right time.

## THE BIRTH OF GENERAL MOTORS

In 1908, the year that the first Model T rolled out, another event occurred that was to have great significance for the automobile industry, for that was the year in which General Motors (GM) was founded. The driving force behind the formation of GM was William C. Durant (1860–1947), a Michigan businessman who had enjoyed considerable success as a manufacturer of wagons. Durant's approach to running an automobile company was radically different from Ford's. Whereas Ford created his company from scratch, Durant built General Motors through acquisitions; as a result of his efforts, GM's stable included Oldsmobile, Buick, and Cadillac, as well as now-forgotten makes like Elmore and Cartercar. Failure to make any money with most of these makes led to Durant's ouster as head of the company in 1910, although he retained a position on the board of directors. Despite his ongoing involvement with General Motors, Durant purchased a number of other automobile firms, one of which had been founded by the well-known Swiss racing driver, Louis Chevrolet. The car's sales successes allowed Durant to trade Chevrolet's stock for shares of General Motors, and in 1916 he was able to regain control of the firm he had founded. With Chevrolet now part of GM's extensive line, the firm produced cars that could be marketed to every economic category of the car-buying public.

## EFFORTS TO EMULATE FORD ABROAD

While holding the premier position in the American automobile industry, the Ford Motor Company extended its influence elsewhere. Ford's principles of car manufacture were first put into practice abroad in 1911 when the company set up an operation to manufacture the Model T in Manchester, England. For many years thereafter, Ford sold more cars in Great Britain than any other manufacturer. Meanwhile, native entrepreneurs tried to emulate Ford's ability to mass-produce cars, although they did not really hit their stride until the mid-1920s. In England, the Morris Cowley, which first appeared in 1915, may be considered that country's first mass-market car. Significantly, its engine, gearbox, rear axle, and steering assembly were obtained from suppliers in the United States until increased tariffs negated the cost advantage of American components.

Mass-market cars designed and produced in Europe differed substantially from the archetypical Model T. The most obvious difference was sheer size. The Model T was not a big car by American standards, but it was substantially larger than the Morris and other cars that aspired to the Model T's level of popularity, as the comparison of the Ford and the Morris in Table 2.1 indicates.

Differences in the sheer size of cars, which persist to this day, reflected substantial differences between the United States and Europe. In the case of the latter, incomes tended to be lower, distances were shorter, roads were narrower, and gasoline was more expensive (due in large measure to higher taxes). Vehicle taxes were often based on horsepower, which was often calculated according to formulas based on an engine's displacement rather than actual measurements of its power. These formulas encouraged manu-

### Table 2.1
### The Ford Model T and the Morris Cowley Compared

|  | Ford Model T | Morris Cowley |
|---|---|---|
| Wheelbase | 8 ft. 4 in. | 7 ft. |
| Track | 4 ft. 9¾ in. | 3 ft. 6 in. |
| Weight | 1,525 lb. | 1,300 lb. |
| Engine Capacity | 2,890 cc | 1,548 cc |
| Length | 11 ft. 2 in. | 10 ft. 6 in.–11 ft. 9 in. |

facturers to produce small-displacement engines, which in turn necessitated the use of low-geared transmissions and differentials to compensate for the engines' lack of torque. Cars built to these standards were usually adequate for European driving conditions, but they were often unsuitable elsewhere.

European automobile manufacturers did not enjoy the economies of scale made possible by a vast car-buying public, precluding complete adoption of the Fordist model of production, in which high capital costs were offset by large production volumes. One indication of the different environment in which European manufacturers operated was their slow adoption of the assembly line. Some manufacturers used conveyers to bring parts to assembly points, and during World War I Renault and Berliet in France, along with Fiat in Italy, used assembly lines on a limited basis, but full-scale assembly lines did not make their appearance until well into the 1920s. The uniqueness of American circumstances meant that the automobile, a German invention that underwent its most significant early development in France, found its largest market by far in the United States. In 1913 the United States had 1,258,000 registered automobiles, one car for every seventy-seven people, while Britain could claim one car for every 165 people, France one for every 318, and Germany one for every 950 (Bardou et al. 1982, 72). In that year, no fewer than 80 percent of the world's automobiles were to be found on the roads of America.

There was nothing uniquely American about the rapid and extensive diffusion of automobile ownership; with the passage of time, what happened in the United States was repeated elsewhere. The benefits of private automobile ownership deeply resonated with people irrespective of their nationality or culture. The Ford saga was eventually replicated in other countries, and a few automobile models eventually eclipsed the Model T's record of more than 15 million cars produced. But many decades went by before this came to pass.

## TECHNOLOGICAL ADVANCES

The quantitative expansion of the automobile was paralleled by substantial qualitative improvements during the opening decades of the twentieth century. Many cars produced at the beginning of the century were truly horseless carriages, high-wheeled vehicles with a one- or two-cylinder engine mounted below the seat. By 1914 the typical car was far more advanced, having a front-mounted engine, usually with four cylinders and side-mounted poppet valves, that drove the rear wheels through a sliding-gear transmission (Ford's two-speed planetary transmission was the great exception). Most of the cars on the road had open bodies because closed bodies with their substantial wooden frameworks incurred a substantial weight penalty. Lighter yet stronger construction came with the all-metal, welded body that was pioneered by the Philadelphia firm headed by Edward G. Budd. This mode of construction also allowed the use of baked enamel finishes, which speeded up the manufacturing process considerably. First used by Hupmobile and Oakland in 1912, all-metal bodies were produced in considerably larger numbers by Dodge in 1914.

Suspensions were usually based on half- or quarter-elliptical leaves. The great majority of cars still lacked front brakes. The rear-wheel brakes were applied through the use of a hand lever, a floor pedal being reserved for an "emergency" brake that worked a drum on the transmission. It was, of course, well understood that this was a less than optimum situation, but speeds were low and motorists encountered far less stop-and-go driving in those days; moreover, front-mounted brakes posed the danger of locking up with wheels that steered the car, with consequent loss of control.

One contribution to making cars safer was the installation of electric headlights in place of lights illuminated by acetylene. Other accessories now taken for granted were notably absent. Windshield wipers did not make their appearance until 1907 when they were fitted to a few French cars; they began to be featured on American cars in 1916. These were operated by hand; wipers powered by electricity or manifold vacuum did not appear until 1919, and were not standard equipment until the 1930s. Brake lights and directional signals were a rarity, as were speedometers and temperature gauges. The first use of a rear-view mirror is often credited to the Marmon that won the first Indianapolis 500 race in 1911; in fact, earlier examples could be found, although many years passed before they were installed as standard equipment on passenger cars.

Tires were the weakest link; one that went as far as 4,000 miles before wearing out was an oddity. Still, their strength was improved when carbon black was added to the rubber in 1912 and a few years later when individual cotton cords replaced cotton fabric in the tire's carcass. Blowouts and flats continued to be frequent occurrences, but detachable rims and wheels eliminated the need for patching inner tubes while the wheel and tire remained on the car. Flat tires were still an annoyance, but now the disabled wheel and tire could be quickly replaced with the "spare."

Virtually all of the automobile's components underwent significant improvements during this period, but it can be fairly argued that of all the technological innovations that occurred during the early years of the twentieth century, none was more significant than the electric starter. Prior to its invention, the most onerous task faced by drivers was manually cranking

Cranking a car to life was inconvenient and occasionally dangerous. Courtesy of the Library of Congress.

the engine to get it started. The typical procedure involved setting the throttle and applying the choke, retarding the spark, and then giving the crank a vigorous swing. If all went well the car would start, but on many occasions numerous turns of the crank were needed to persuade the engine to come to life. Even worse, there always lurked the danger that the engine would briskly rotate in the opposite direction, and the operator's arm could be broken by the violent movement of the crank.

One such incident turned tragic when a young automobile executive's jaw was broken when the crank of a car he was trying to start slammed into his face. Gangrene set in, and he died a few days later. This tragedy motivated Henry Leland (1843–1932), Cadillac's general manager, to seek an effective method of starting a car with no physical effort and at a safe distance. Previous attempts at self-starters using compressed air or springs had been only marginally successful, and inventors attracted to electric motors were put off by calculations that seemed to indicate that an effective starter motor would have to be about as big as the engine itself. This apparent impasse was solved by Charles Kettering (1876–1958), an engineer at Cadillac who previously had worked with electrically powered devices while employed at the National Cash Register Company. Kettering's experience told him that, just as with a cash register, a car's starter motor only had to run for a few seconds at a time, which allowed it to be much smaller than a motor under constant load. Kettering's efforts produced the first effective self-starter, which made its appearance on the 1912 Cadillac.

The electric starter removed one of the chief obstacles to car ownership for many people, and was of particular importance in expanding the number of women motorists. Shifting gears was still a challenge to many, but at least the danger of a broken arm or worse had been eliminated. The self-starter also indirectly reduced the skill required to drive a car by allowing cars to be powered by larger engines with higher compression ratios; these produced more torque and thereby lessened the need for frequent gear shifting. The self-starter also rendered irrelevant one of the chief virtues of the electric car; already suffering from falling sales, it all but disappeared in the years that followed.

## ROADS FOR CARS TO TRAVEL ON

An important selling point for the Model T Ford was that it sat high off the ground. This gave it an ungainly appearance, but it was a useful attribute in a driving environment notably lacking in well-prepared roads. As was noted earlier, a large segment of the American population lived in the countryside in the Model T's heyday, and farmers were some of the car's most important customers. Although Model Ts could be used for agricultural tasks such as plowing and threshing, the real significance of the Model T and other cars is that they gave people the opportunity to break out of the isolation that had been an inherent feature of rural America. For urbanites, the automobile promised at least a temporary escape from the city and a chance to spend some time under clear skies in bucolic settings.

That at least was the promise, but for city and country folk alike, the ability to drive somewhere was limited by the sorry condition of America's roads. Towns and cities could boast having some paved streets, but what passed for highways between population centers were largely dirt roads that were impassable when the snow fell, turned into bogs when it rained, and exposed the drivers of cars and other vehicles to clouds of dust during dry periods.

Technologies for constructing all-weather roads were developed in the late eighteenth and early nineteenth centuries by Pierre-Marie Trésaguet (1716–1796) in France and Thomas Telford (1757–1834) and John Louden McAdam (1756–1836) in Britain. Their techniques involved the use of layers of stone arranged in such a way as to promote drainage. The road surfaces were generally unpaved, although asphalt (a heavy petroleum constituent, or "fraction," known as bitumen that is mixed with sand and bits of stone) was occasionally used in France beginning in the mid-nineteenth century. Concrete was used to bind together stones of macadam roads as early as 1865, but not on a widespread basis. These methods of road building were adequate for horse-drawn carts, but they were not up to the task as far as automobile traffic was concerned. The efficient operation of automobiles necessitated paved roads, but these were a distinct minority; in 1909 fewer than 9 percent of the roads in the United States had any kind of surfacing, and the countryside was completely lacking in paved roads (Flink 1988, 169). Asphalt came to be the most common road surface, but as late as 1913 only four states had more than 1,000 miles of roads (other than urban streets) paved with asphalt (McShane 1994, 219).

Road construction languished due to an unwillingness to spend time, money, and energy on highway improvement. This situation began to change in the late nineteenth century due to widespread ownership of a vehicle that in numerous ways prepared the world for the automobile—the bicycle. As with the early days of the automobile, the bicycle's appeal was more sporting than utilitarian. Cyclists' clubs organized group rides, but their duration was limited by the lack of effective roads. In 1892 a national bicyclists' organization, the League of American Wheelmen, combined with a farmers' association known as the National Grange of the Patrons of Husbandry to establish the National League for Good Roads. The league was not alone in its enthusiasm for road improvement; even the railroads came around to supporting better roads, under the fatally flawed assumption that they would facilitate the movement of passengers to railroad stations.

Bicyclists and farmers were joined by growing ranks of motorists clamoring for the public financing of road construction and maintenance. Roads had historically been financed by local governments or through tolls that were collected for both public and privately owned highways. But the rapid expansion of automobile travel allowed a new form of taxation: the gasoline tax. The collection of taxes is hardly the most popular activity of governments, but in the United States, gasoline taxes, directly tied as they were to road building and upkeep, met with widespread approval. In 1919 Oregon, New Mexico, and Colorado became the first states to levy a tax on gasoline. In the next decade, every state instituted a gasoline tax (typically 3 or 4 cents per gallon) and used the revenues as the prime means of funding road construction and maintenance. Unlike other taxes, the gasoline tax met with very little resistance. Because it was directly tied to road expenditures, it was welcomed for its contribution to fast and comfortable automobile travel. As one Tennessee tax official mused, "Who ever heard, before, of a popular tax?" (Flink 1988, 171).

Most other countries took a different view of gasoline taxation. Rather than being collected exclusively for roads and other components of automotive infrastructure, gasoline taxes were treated simply as a source of revenue to be used for whatever purposes the government deemed appropriate. And given many governments' insatiable desire for revenues, gasoline taxes eventually accounted for a large share of the price of a tankful of gasoline. In turn, differences in the price of gasoline had a powerful effect on the design and size of cars, resulting in European cars being for the most part smaller and more fuel efficient than their American counterparts.

## A PATENT ON THE AUTOMOBILE

The early history of the automobile in the United States is one of intense competition, as hundreds of firms went into the business of car manufacture. Most of them failed in short order, but their place was taken by legions of eager newcomers. Yet amidst this competitive scramble, the industry was gripped by a successful attempt to wring monopoly profits from the industry. The story began in 1876 at the Philadelphia Centennial Exhibition when George B. Selden (1846–1922), a successful attorney, witnessed the operation of a two-stroke engine that had been patented by George Brayton four years earlier. It enjoyed some success as an industrial engine, and Selden thought it could be used to power a vehicle. Accordingly, in 1879 he filed a patent application for "an improved road engine" powered by a "liquid hydrocarbon engine of the compression type." Selden had no intention of actually building a car, and he delayed the actual issuance of the patent until 1895, by which time a number of firms were in the business of manufacturing automobiles. As holder of the patent, Selden had the legal right to collect a royalty for every car sold, but the costs of enforcement would have been substantial, so in 1899 he sold the patent to a financial syndicate known as the Columbia & Electric Vehicle Company (later simply the Electric Vehicle Company), at the same time retaining the right to collect a portion of future royalties. A year later the Electric Vehicle Company began to press patent infringement suits, and within a few years it was able to use its ownership of the patent to create a legal monopoly; selected firms were licensed to use the Selden patent in exchange for royalty payments, while others were barred from car manufacture altogether. Beginning in 1903, members of the patent pool were joined in the Association of Licensed Automobile Manufacturers (ALAM). Firms outside the association, along with the people who bought their products, faced the threat of expensive lawsuits.

One manufacturer whose application to join the ALAM was rejected was Henry Ford. Ford went on making cars, and was found guilty of patent infringement after a six-year court battle that commenced in 1903. The ruling was overturned on appeal in 1911 when three circuit court judges ruled that although the Selden patent was valid, it only covered cars powered by the long-extinct Brayton engine. Ford's successful challenge of the Selden patent won him considerable acclaim as a foe of "monopolists" and an advocate for the common man, even though he was on his way to becoming one of the richest men in the United States. The case also demonstrated how the ownership of a key patent can pose a serious threat to an entire industry. In recognition of this, most American automobile manufacturers subsequently entered into cross-licensing arrangements that gave each firm the right to use inventions patented by other firms in the industry (Greenleaf 1961).

## FUELING THE AUTOMOBILE

As the previous chapter noted, by the middle of the first decade of the twentieth century the internal combustion engine had emerged as the dominant means of powering the automobile. Although some internal combustion engines were designed to use kerosene, the most common fuel was gasoline, a combination of liquid hydrocarbons that emerges when crude oil is refined. Oil refining became an important industry during the latter third of the nineteenth century, but its objective was the production of kerosene for heating and illumination. Gasoline was a more dangerous substance due to its volatility and extreme flammability, and often was treated as an undesirable byproduct of petroleum refining. It was sold as a fuel for stationary farm and industrial engines, but much of it was dumped into the nearest body of water due to lack of demand.

All this changed with the rise of the automobile. Although gas used for illumination had fueled early stationary internal combustion engines, gasoline was far better suited to the powering of automobiles, as it packed a great deal of energy in a small volume. Thus, gasoline had become a valued commodity as rapidly increasing numbers of cars took to the road. The key question now was whether there would be sufficient supplies of it to fuel an expanding automotive fleet. The world's known stock of petroleum had expanded significantly with the discovery of extensive fields in East Texas

By the 1920s service stations and gasoline pumps were common roadside features. Courtesy of the Library of Congress.

in 1901, but many observers were of the opinion that petroleum reserves would be insufficient in the long run. As was to happen when the supply of petroleum became problematic many decades later, alcohol was proffered as an eventual substitute for petroleum-based fuels.

In fact, the ability to extract sufficient quantities of crude oil never was a limiting factor in the spread of automobile ownership during most of the twentieth century. Much more problematic was refining the oil into gasoline, and then getting the gasoline to the customer. At first, gasoline was obtained through distillation; as petroleum was heated, its various fractions vaporized in accordance with their different boiling points. Gasoline with its low boiling point was one of the first fractions to be separated from the petroleum; it then passed through a condenser, where it returned to liquid form. This process had its limitations, as only 15 percent of the petroleum could be converted to gasoline. This limitation was surmounted by William Burton's (1865–1954) development of thermal cracking, through which high temperatures and pressures were used to break large hydrocarbon molecules into smaller ones. First employed in 1913, it doubled the amount of gasoline that could be obtained from a given quantity of petroleum. Even more gasoline was obtained through the use of catalytic cracking, invented in France by Eugene Houdry (1892–1962). When it began to be used in the United States during the mid-1930s, the yield of gasoline increased to 43 percent.

Although the first facility devoted to the fueling of automobiles opened in Bordeaux, France, in 1896, early motorists usually filled their tanks at general stores, repair garages, and the like. In the United States, the first drive-in gas station opened in St. Louis in 1905, and the first station expressly designed to dispense gasoline went into operation in Detroit in 1911. At first, it pumped 100 gallons per day; six months later its daily sales had increased twentyfold (Anderson 1984, 279). Gasoline stations and other roadside services proliferated within the next few years, and their appearance in cities, towns, and the countryside began to reflect the demands of automotive travel.

## WOMEN AT THE WHEEL

The automobile often has been viewed as a "masculine" object, but this should not obscure the fact that women have been involved with cars from the very beginning. We have already seen how Bertha Benz can be credited with taking the world's first long-distance car trip. But the great majority of women, like most men, did not have access to an automobile until much later. If a woman wanted to go any distance within a town or city, she had to rely on public conveyances like horse-drawn or electric streetcars. Trips of this sort could be particularly unpleasant for women, who often were subjected to the unwanted attentions of obnoxious male passengers. Under these circumstances, the opportunity to get around in a private car was particularly appealing.

Wealthy women, like their male counterparts, had the option of employing a chauffeur for automotive transport, but even when they could afford the services of a professional driver, they often preferred to take their place behind the wheel. The admission of women to the ranks of the motoring public did not come easy. Many men were of the opinion that women lacked the physical strength, intellectual capacity, and emotional stability to drive a car, but these attitudes were difficult to maintain as increasing numbers of women joined the ranks of automobile owners and drivers. Although women comprised a small percentage of early motorists, some of them distinguished themselves by engaging in long-distance car travel, and some even competed in races. Although it was headed by a man who was anything but progressive in his social philosophy, the Ford Motor Company recognized the growing significance of automobiles in the lives of women. One early Ford publication proclaimed that the automobile "has broadened her horizon—increased her pleasures—given new vigor to her body—made neighbors of faraway friends—and multiplied tremendously her range of activity. It is a real weapon in the changing order. More than any other—the Ford is a woman's car" (quoted in Scharff 1991, 54).

Although women drivers had ceased to be a novelty by the end of the twentieth century's first decade, the notion that men and women inhabited "separate spheres" persisted. Although men might grudgingly share the road with women motorists, this was often coupled with the belief that women were much better suited to the operation of low-performance vehicles. In this schema, cars powered by internal combustion engines were the domain of men, while electric cars were thought to be particularly well suited to women drivers. Under these circumstances, a key weakness of the electric car, its lack of range, could even be seen as a virtue because it kept women within a comfortable radius of their homes. It soon became apparent, however, that women wanted the same things from their cars that men did, and the limitations of the electric prevented it from gaining wide acceptance among women motorists.

## SOCIAL CHANGE AND THE AUTOMOBILE

The incorporation of women into the ranks of motorists was part of a larger process in which commonly accepted social arrangements were being reshaped by automobiles and the people who owned and drove them. At the beginning of the twentieth century, automobiles were few in number and their influence on society and culture was minimal, yet within a few years significant changes in the physical and social environment reflected the growing influence of the automobile.

The automobile was a creation of industrial society, and cities were its initial domain; in 1910 urban residents were four times more likely to own a car than rural residents (McShane 1994, 105). Even so, it can be fairly argued that the automobile's influence was most strongly felt in the countryside, especially in the United States. Early motorists occasionally encountered hostility from farmers whose horses had been spooked or whose lives were otherwise disrupted by automobiles invading their territory, but within a few years rural America had come to appreciate the advantages of automobile ownership. In 1911 the U.S. Department of Agriculture counted about 85,000 cars on American farms; by 1920 that figure had increased over thirteen times to nearly 1,150,000. In that year, 30.7 percent of farms included an automobile (Berger 1979, 51–52).

Much of the automobile's appeal lay in its ability to free rural people from the physical and cultural isolation that was a characteristic feature of life in the countryside. The operating radius of a horse and buggy was no more than twenty miles, and this distance effectively defined the world that rural people inhabited. Within its confines people did their shopping, attended church services, sent their children to school, engaged in leisure pursuits, and even found marriage partners. Automobile ownership transformed the countryside by expanding the spatial area in which rural life was lived. Automobile owners could travel to large towns and even metropolitan centers for their shopping. Children could be transported to large consolidated schools that began to replace schools that had packed students of all ages into one or two rooms. Worshipers were no longer limited to a small village church served by a circuit-riding minister. Car excursions offered novel recreational opportunities, not all of which were in accordance with traditional moral values. In sum, car ownership significantly narrowed the age-old gulf between urban and rural life.

With greater mobility, however, came the attenuation of social and economic ties within local communities. Shops, churches, schools, and physicians that had served a captive clientele now had to compete with a large number of competitors within driving distance, and many of them were not up to the task. The general store, the country church with its miniscule congregation, and the one-room schoolhouse began to fade from the scene, undermining the institutional basis of the tightly knit rural community. To be sure, the automobile was not the sole cause of these changes; contemporaneous with or predating the automobile were mail-order retailers, movie theatres, and mass-circulation magazines delivered through rural free delivery that expanded the horizons of rural people. But there is no substitute for physically getting out of a familiar environment and going to another one, and no conveyance did that better than the private car.

Urban life also began to be transformed as a result of growing automobile ownership. One prominent benefit of automotive transportation was the reduction of the "exhaust" left behind by large numbers of horses. In New York City toward the end of the nineteenth century, horses deposited around a million pounds of solid excrement every day (McShane 1994, 51), creating an enormous disposal problem that was exacerbated by the need to remove the carcasses of horses that died in the streets. Horse manure also posed a serious public health threat; billions of flies bred in it, and dry, pulverized manure blew everywhere, carrying with it pathogens responsible for tetanus and tuberculosis, as well as a variety of other respiratory diseases.

The piling up of horse manure in the streets wasn't the only unpleasant aspect of urban life, as overcrowding, air pollution, criminal activity, and traffic conspired to make daily life in the city a struggle. The desire to leave urban ills behind was especially strong in the United States, where a Jeffersonian belief in the morally suspect nature of city life was coupled with medical theories that ascribed a variety of health problems to overcrowded living situations. While many urbanites sought a more rural lifestyle, the majority of offices and factories remained within the central city, so transportation between home and work became a necessity for people who had taken up residence in the suburbs. At first, this need was met by extensive networks of trolley lines, but they confined suburban development to areas close to the tracks. Commuting by automobile removed this restriction and opened up new tracts of land for suburban development. As early as 1906 it was noted that "places fortunately situated, and blessed with healthful surroundings, which formerly were undesirable because of their inaccessibility, are rapidly developing, according to the local real estate authorities, who are free to admit that they mention the automobile as the means of transportation when soliciting the patronage of their purchasers" (quoted in Flink 1970, 109).

In addition to making more land available for housing, the automobile offered commuters privacy and flexible travel schedules. According to some of its enthusiasts, the automobile promised an idyllic lifestyle that combined urban work opportunities with a rural lifestyle; as one journalist rhapsodized in 1904, "Imagine a healthier race of workingmen . . . who, in the late afternoon, glide away in their own comfortable vehicles to their little farms or homes in the country or by the sea twenty or thirty miles distant! They will be healthier, happier, more intelligent and self-respecting citizens because of the chance to live among the meadows and flowers of the country instead of in crowded city streets" (quoted in Flink 1970, 109).

The automobile's appeal was not limited to the practical advantages it brought to farmers, city-dwellers, and suburbanites. Motoring was touted as beneficial to one's health, sometimes to an absurd degree. One journalist went so far as to claim in 1903 that the automobile "is the greatest health giving invention in a thousand years. The cubic feet of fresh air that are literally forced into one while automobiling rehabilitate worn-out nerves and drive out worry, insomnia, and indigestion. It will renew the life and youth of the overworked man or woman, and will make the thin fat and the fat— but I forbear" (quoted in Flink 1970, 106).

Cars also offered a level of visceral excitement rarely experienced before. Although by the end of the nineteenth century a few express trains were reaching speeds in excess of 100 mph, the automobile gave the ability to attain high speeds to the man or woman at the wheel. As one journalist noted, "The sensation which arouses enthusiasm for the automobile comes almost solely from the introduction of the superlative degree of speed and from the absence of effort or fatigue. The automobile is a vehicle that touches a sympathetic chord in all of us" (quoted in Flink 1970, 100–101). Others went even further, finding unparalleled aesthetic appeal in a speeding automobile. As Filippo Marinetti, a young Italian artist, asserted in "The Foundation and Manifesto of Futurism," "We declare that the world's splendor has been enriched by a new beauty: the beauty of speed. A racing car with its hood adorned with great pipes, like serpents of explosive breath—a roaring car that seems to be operating like a machine gun, is more beautiful than the Winged Victory of Samothrace" (1909, quoted in Carrieri 1963, 12).

Not every driver was intoxicated with speed, but many took pleasure in the automobile's ability to fulfill another common human quality, the desire to flaunt one's elevated social status. The cost of buying and operating early automobiles largely confined them to the well-to-do, or least the upper reaches of the middle class. Consequently, early motorists aroused the ire of the poorer, nonmotoring public. This hostility was muted as automobile ownership rapidly expanded, and an emerging motoring public began to indulge in the multifaceted delights of driving. But, as we shall see, the appeal of the car as a status symbol did not disappear as automobile ownership spread and practical utility came to be taken for granted.

During the first years of the twentieth century, the automobile ceased to be a mechanical oddity and was well on the way toward being a key artifact of the new century. In 1914 the automobile market in America seemed to have no limits, while in Europe the mass production of automobiles was building up momentum. Tragically, however, the economic advance that was both cause and effect of the expansion of automobile manufacturing and ownership came to a sudden halt as the industrialized world plunged into the abyss of World War I. Yet for all the damage and human suffering it caused, the war proved to be only a temporary setback in the advance of mass automobility. The adolescent was growing rapidly, and it would reach a certain level of maturity in the years to come. Maturity, however, did not assure the smooth assimilation of the automobile into society; instead, the growing numbers of automobiles and their drivers would give rise to a new set of issues that were not easily resolved.

# 3

# From Battlefield to Boulevard, 1914–1929

◆

During the first decade of the twentieth century, the automobile ceased to be a novelty. Car ownership expanded rapidly, and was only temporarily sidelined by the most destructive war in human history up to that time. In the years that followed World War I, production expanded almost every year, and automotive technology advanced on many fronts. The industry's dominant firm began to fade as others moved to the front. And all the while, the automobile had indisputably become one of the most important sources of social and cultural change at home and abroad.

## THE AUTOMOBILE GOES TO WAR

Very early in its history, the automobile had proved its mettle under exceptionally difficult conditions. In the aftermath of the San Francisco earthquake of 1906, fires raged throughout the city, public services broke down, and rubble was strewn everywhere. Under these adverse circumstances, automobiles demonstrated their utility and reliability as they evacuated the injured and brought in medicine, food, and supplies, all the while surpassing the performance of horses and mules even though they sometimes were forced to drive on bare wheels after infernally hot streets melted their tires.

The manmade conflagration of World War I, or the Great War as it was known at the time, further demonstrated the capabilities of the automobile. Cars were in the spotlight soon after hostilities began in 1914, when a fleet of Parisian taxicabs conveyed troop reinforcements during the Battle of the Marne, helping to blunt a German invasion that could have resulted in a catastrophic defeat for the French. Less dramatically but more significantly over the long run, cars performed crucial duties in sustaining the war effort. Although mules and horses continued to be used extensively, automobiles and other motorized vehicles performed essential roles in transportation, scouting, courier duties, ambulance services, and other war-related tasks.

The majority of automobile drivers were men, but the war also created an unprecedented opportunity for women to serve as drivers in war zones as well as on the home front. Wartime demands and a shortage of men undermined traditional expectations about proper feminine activities as significant numbers of women drove, maintained, and repaired cars and trucks. In doing their share for the war effort, women delivered supplies, transported wounded soldiers, and ferried off-duty servicemen. This recasting of gender roles set the stage for even greater changes that were to come after the war in regard to the operation of motor vehicles and many other things as well.

Although many different makes of cars were pressed into military service, once again the Model T was most prominent, with 125,000 of them used by the Allied forces. Battlefield conditions had some similarities to the rural environment, and the Ford's ability to travel over poor or nonexistent roads, along with its durability and ease of repair, demonstrated the essential correctness of its design.

In addition to supplying cars, the automobile industries of the belligerents were heavily involved in war production. Trucks were an obvious need, and their production in the United States doubled during the last year of the war. The manufacturing capabilities of automobile manufacturers also were employed for the production of shells, guns and gun carriages, tractors, helmets, and many other articles of war. The production of aircraft engines was particularly well suited to the skills and equipment of the automobile industry, and firms on both sides were prominent in this sector. In Germany, Daimler and Benz (separate firms until their merger in 1926) were key suppliers, while Hispano-Suiza engines powered airplanes flown by their French adversaries. In England, Rolls-Royce took up the manufacture of airplane engines, a line of work that would eventually outstrip luxury car manufacture as the firm's largest source of income and profit.

One of the most important contributions of American industry to the war effort was the Liberty aircraft engine. Reportedly designed by two automotive engineers in only three days in a Washington hotel room in 1917, the Liberty was expressly designed to be manufactured through the use of American mass-production techniques. Nearly 25,000 of them were built, almost all of them by automobile firms. In all, 30,000 of the 42,000 aircraft engines built in the United States during the war were turned out by American automobile manufacturers (Rae 1965, 73). A desire to devote his efforts to aircraft engine production caused Henry Leland, the chief executive at Cadillac, to leave that firm in order to found the Lincoln Motor Company, which went on to build 6,500 Liberty engines before turning to the manufacture of luxury automobiles after the war (Rae 1965, 80).

World War I did not seriously harm the American automobile industry even though the government restricted the supply of steel to the industry in 1918, substantially cutting into production and profits. Henry Ford initially stated that he would forego any profits earned through wartime military sales; this patriotic gesture was subsequently forgotten, but his firm did suffer an erosion of profits as car sales languished as a result of decreased production. European manufacturers never had the sales volumes of American firms, and in many cases the shift to wartime production actually increased the scale of their operations. In England, employment at Austin's main factory shot up from 2,300 in 1914 to 20,000 in 1919 in response to military orders placed by the Ministry of Munitions. Fiat became the third largest company in Italy largely on the strength of its wartime orders, and Mercedes was able to increase its capital fourfold (Flink 1988, 76).

## THE EMERGENCE OF THE TRUCKING INDUSTRY

In addition to providing a financial windfall for some automobile firms, World War I gave an important stimulus to the emergence of the long-distance trucking industry. In an age of industrialized warfare, military outcomes have hinged on a nation's ability to promptly supply sufficient quantities of armaments, ammunition, uniforms, rations, and other accoutrements of war to fighting forces. In its scale and scope, World War I exceeded all other wars that had come before it, requiring the delivery of enormous amounts of materiel to its combatants. The war also produced massive numbers of casualties, so motorized ambulances played an essential role in evacuating troops that had suffered horrible onslaughts of machine gun fire, artillery assault, and poison gas attack.

The automobile industry responded rapidly to meet the military's transport needs by massively increasing truck production. The conversion from car to truck production posed few technical challenges because car chassis could be readily adopted to carry truck bodywork. Under the stimulus of wartime demands, U.S. truck manufacturing mushroomed from 24,900 vehicles in 1914 to 128,000 in 1917 (Flink 1988, 78), and as early as 1915 the industry was producing more trucks in a month than it had in any past year (Wren and Wren 1979, 62).

The transportation needs of World War I forced a greater reliance on trucks for overland transportation. The railroads had been hard-pressed to meet the demands of wartime transportation, which included shipping trucks made in the Midwest to the ports on the East Coast from which they embarked on their voyage to Europe. To alleviate the strain on the railroads, Roy D. Chapin, the head of the Highway Transport Committee of the Council for National Defense, arranged for convoys of trucks to make the journey under their own power while carrying a load of freight at the same time. In all, 18,000 trucks made the trek from the end of 1917 to the 1918 Armistice. This use of trucks for long-distance haulage relieved some of the pressure on the overburdened railroad system, and equally important, it demonstrated that truck transportation could be reliable and economical. All in all, truck transportation played a significant role in the winning of the ghastly war of attrition that was World War I. As a member of Britain's war cabinet noted in 1919, the Allies had "floated to victory in a wave of oil" (quoted in Flink 1988, 73).

## ROAD IMPROVEMENTS

In the years immediately following the cessation of hostilities, the military continued to explore the potential of motorized transportation. The performance of trucks during World War I convinced General John Pershing, who had served as commander of U.S. troops in Europe, to organize a cross-country convoy of seventy-nine trucks that crossed the United States from Washington to San Francisco in 1919. Hampered by the poor state of the roads upon which they traveled, the trucks averaged only fifty miles a day, and needed fifty-six days to complete their journey (Lay 1992, 172).

At this point the difficult road conditions experienced by the U.S. Army had been only modestly addressed by the federal government because roads were largely the domain of state and local governments. The federal government's involvement in road improvement began with a 1912 appropriation act that provided some funding for rural roads used for mail delivery. This piece of legislation was augmented by the 1916 Federal Aid Road

Early motorists often had to cope with impassable roads. Courtesy of California Department of Transportation.

Act, which empowered the secretary of agriculture to engage in road improvements that would help farmers bring their produce to market. In 1921 the federal government finally got serious about the miserable state of the nation's roads through the passage of the Federal Highway Act. As stipulated by the act, highway departments of each state could designate up to 7 percent of their nonurban roads as "primary"; these roads then could receive federal matching funds on a 50-50 basis. Due in part to this initiative, considerable progress was made in the 1920s as federal, state, and local authorities spent about $2 billion a year on roads and highways.

Government assistance put some American roads on par with the roads of some European countries. In the meantime, however, road technology was moving ahead on the other side of the Atlantic. Italy led the way with the construction of its first *autostrada*, a limited-access, divided road that linked Milan with Varese and Lake Como, a distance of about 30 miles (50 km). Completed in 1925, the *autostrada* was designed for speeds of 36 mph (60 kph), a rapid clip in those days. Primarily intended to serve as a showpiece for the "New Roman Empire" of Benito Mussolini, the *autostrada* was of limited importance at a time when very few Italians owned cars. Even so, it can be considered a forerunner to the German autobahns and American freeways that appeared during the following decade.

## THE INTERCITY BUS

In addition to giving a boost to car and truck transportation, better roads stimulated the emergence of a new mode of long-distance passenger travel, the intercity bus. Long-distance bus travel began modestly during the 1910s when local entrepreneurs used their cars to shuttle a few paying passengers to their destinations. These operations lacked fixed schedules and routes, and passengers were usually picked up and dropped off at curbside. Early bus services of this sort were popularly known as "jitneys," which at the time was the slang expression for a nickel, the most common cost of a ride. Most of these services were confined to a single community, but a few intercity lines with regular schedules went into operation at this time.

As the industry matured, specially built motor coaches replaced the elongated automobiles of the recent past. The first of these was built in California by Fageol in 1921, and other manufacturers, some of them specializing in bus manufacture, soon followed. By 1928 it was even possible to take a coast-to-coast bus ride, a journey that took five days and 14 hours and entailed 132 stops along its route from Los Angeles to New York (Walsh 2000, 21).

As long-distance bus service expanded, large firms with far-flung route systems became the dominant players in the industry. Like many other segments of the automobile industry, economies of scale stimulated the formation of large firms. The most long-lasting and successful of these was Greyhound, which originated as a number of separate firms in the Upper Midwest that were amalgamated and expanded into a nationwide system during the late 1920s.

By this time, travel by intercity bus had become a significant part of the American scene, with nearly 21,500 vehicles traveling 218,601 route miles (Walsh 2000, 8). Private automobiles were now the dominant way of getting from one place to another, so buses accounted for only 3.3 percent of intercity travel, but 17.4 percent of public-carrier passenger miles (Walsh 2000, 27). Intercity buses did not threaten the rapidly growing use of automobiles for long-distance travel, but they added to the woes of the nation's railroads by cutting into their passenger market.

## DECLINE FOLLOWS TRIUMPH AT THE FORD MOTOR COMPANY

In 1923 the Ford Motor Company produced 1,817,891 Model Ts, more than in any previous year. It also marked the peak of the car's production.

Although Ford appeared to hold an insurmountable position among the world's automakers, it was in fact on the threshold of a decline from which it would never completely recover. The success of the Model T had been based on relentlessly driving down production costs so the car could be sold at rock-bottom prices. Between 1921 and 1925, Ford cut prices six times, resulting in a new Model T runabout carrying a price tag of only $260 in 1925. But automotive design and technology had not been stagnant, and customers' expectations had been steadily rising. The Model T was tough and reliable, if a bit quirky, but despite substantial upgrading, such as the fitting of an electric starter, it remained a design from the century's first decade. It had been a perfect car for its time, but times were changing. For a few hundred dollars more, prospective customers could choose from a wide array of cars with more powerful engines; greatly improved steering, suspension, and brakes; higher levels of comfort; and a generally more modern appearance. At the lower end of the scale, would-be purchasers could benefit from a new automotive phenomenon: the massive expansion of the used-car market. A serviceable Model T could be acquired for $50 or less, and for the price of a new Model T it was possible to buy a late-model used car that surpassed the Model T in every respect. The Tin Lizzie had been so successful that it constituted half of the cars on American roads, but many of their owners were ready to exchange them for something better.

By 1926 the share of American automobile sales taken by the Model T had dropped to 30 percent (Hounshell 1984, 263–64), and even Henry Ford had to accept that something more modern was needed. The introduction of the Model T's replacement necessitated a complete overhaul of Ford's manufacturing operations. On May 26, 1927, the last of the 15,007,033 Model Ts made since 1908 rolled off the assembly line. Ford's main factory then shut down, and six months elapsed while the firm prepared for the production of an all-new Ford, the Model A, which was shown to the public at the end of the year and went on sale in early 1928.

The car that Ford's fate now rested on was a well-engineered car with none of the eccentricities of the Model T. It was powered by a 200.5-cubic-inch, 40-hp, four-cylinder engine connected to a conventional three-speed gearbox. The planetary transmission was a thing of the past, although ironically the principle was revived for the first effective automatic transmission that emerged in the late 1930s. Like most of its contemporaries, the Model A had solid-axle front suspension and brakes on the front wheels, although their mechanical actuation reflected Henry Ford's distrust of hydraulic brakes, probably because he could never fathom how they worked. Its selling price of $430 to $695, depending on the model, made it a far better value than the Model T it replaced. All in all, the Model A was a good

Ford's River Rouge plant was one of the world's largest industrial complexes, complete with its own steel mill. Courtesy of the Library of Congress.

car, the virtues of its basic design reflected in the fact that tens of thousands are still in operation today.

The length of the shutdown required to tool up for the Model A underscored a key drawback of Fordist methods of manufacture: the standardization of the product and the methods by which it was built had made it extremely difficult, time-consuming, and costly to deviate from a set path. The Model A briefly allowed Ford to regain many sales previously lost to other manufacturers, notably General Motors, but the firm's days of dominance were over.

## GENERAL MOTORS PASSES FORD

The rise of General Motors neatly parallels the relative decline of the Ford Motor Co. Founded in 1908, the same year that Ford rolled out the first Model T, General Motors came to exemplify a way of doing business that was the antithesis of Ford's. The differences began with the men who created each firm. Henry Ford was above all concerned with expanding the market for his cars through a never-ending quest for greater production efficiency. GM's founder, William C. Durant (1861–1947), was not particularly interested in the design or manufacture of cars; his talents lay in perceiving where the market was going and producing cars that would best satisfy that market.

In pursuing this strategy, Durant had the advantage of having Buick and Cadillac in GM's stable, but the corporation also had a number of car lines that failed to appeal to prospective buyers. As noted in the previous chapter, dissatisfaction with Durant's leadership led to his ouster in 1910. At this point, most of GM's component enterprises were running at a loss, and GM survived only through the financial assistance of a banking syndicate that took over management of the firm. Charles W. Nash (1864–1948), Buick's top executive, became president of GM in 1912, while Walter P. Chrysler (1875–1940) took over for him at Buick. In the meantime Durant had formed a company to manufacture a low-priced car named after Louis Chevrolet, a well-known Franco-Swiss racing driver. The car was a sales success, and by swapping five shares of Chevrolet stock for each General Motors share, Durant was able to regain the presidency of GM in 1916. GM enjoyed some good years, but car sales began to plummet during the sharp postwar recession that began in 1920. By this time several of his most capable managers, notably Nash and Chrysler, had tired of Durant's erratic managerial style. They left GM and eventually went on to found the successful automobile companies that bore their names.

Alienating key employees was not Durant's only managerial shortcoming. Durant was heavily involved in stock speculation, and his frenetic buying and selling of stocks was a serious distraction from his leadership of GM. Moreover, GM's major shareholders feared that if Durant suffered a personal bankruptcy, it would seriously erode public perception of the company he headed. Accordingly, the DuPont Company, already a major holder of GM stock, bought 2.5 million additional shares with the proviso that Durant once again relinquish the presidency. Durant went on to create a car company named after himself, but despite some initial success it folded in less than ten years.

Pierre S. du Pont (1870–1954) took over as president of GM, but he left the running of the firm to his executive vice president, Alfred P. Sloan Jr. (1875–1966). Sloan had come to GM when the company he headed, the Hyatt Roller Bearing Company, was acquired during Durant's second reign as GM chief. The firm had been founded by John W. Hyatt (1837–1920), whose tapered roller bearings were widely employed within the automobile industry. Hyatt had not been interested in day-to-day management, and presidency of the firm had passed to Sloan. Like many other GM executives,

Sloan had been frustrated by Durant's managerial style, and in 1920 he was about to quit when Durant's ouster was followed by the opportunity to become president of General Motors.

Upon taking the presidency of GM, Sloan began to put into practice a reorganization plan that he had originally presented to Durant, only to have it ignored by the GM chief. The firm that Sloan took over was a sprawling hodgepodge of firms, but Sloan saw a strength in all this diversity, and instead of bringing everything under centralized control, he created a kind of manufacturing federation. Individual car brands like Chevrolet and Buick would be constituted as semiautonomous divisions with considerable latitude in the design, engineering, manufacture, and marketing of their products. At the same time, they would be subject to central control exercised by corporation-wide executive and financial committees. This organizational structure allowed divisional heads to become intimately familiar with their products, customers, and suppliers, and in so doing produce the cars that would have the greatest appeal within their segment. At the same time, the divisional structure reinforced Sloan's belief that GM should make cars to suit "every purse and purpose." This created a natural purchasing hierarchy within GM; a customer might start off buying a Chevrolet, and then move up to an Oldsmobile as his or her economic fortunes advanced. At the upper reaches of the product hierarchy could be found Buicks and Cadillacs for those who enjoyed above-average financial success.

Sloan also introduced a marketing ploy that became standard procedure in the automobile industry for many decades to come: the annual model year. Although it has never been true that automobile manufacturers have attempted to insure future sales by deliberately designing cars to fall apart after a few years, General Motors made heavy use of "planned obsolescence" in order to induce people to trade in their cars for new models. As implemented by the firm from 1923 onward, although a car's basic chassis and running gear remained largely unchanged for a number of years, the last few months of each calendar year would see the introduction of an "all-new" model in which only the design of the body and interior had been altered. The new model might not be a substantive improvement over its predecessor, but when the marketing was effective, millions of consumers could be induced into jettisoning perfectly serviceable cars that bore the stigma of being stylistically obsolete.

GM's emphasis on style was reflected and reinforced by its establishment in 1927 of the first automotive design studio employed by a major manufacturer, the Art and Colour Section. Led by Harley Earl (1893–1969), who had gotten his start designing custom bodies for ultraexpensive chassis supplied by Duesenberg and the like, the Art and Colour Section introduced such innovations as the use of full-size clay models for designs in progress, which helped stylists to envisage the smoother, rounder shapes that began to characterize body design. Earl's influence eventually spread throughout the entire industry as competing firms hired designers who worked under Earl to staff their own styling departments. Virtually every automaker adopted Earl's stated philosophy of automotive design, which was "to lengthen and lower the American automobile, at times in reality and always at least in appearance" (Flink 1988, 236).

GM's style leadership was also enhanced by its early lead in the marketing of attractively painted mass-market cars. At the time that Durant retook control of GM, his main financial backer was Pierre S. du Pont, whose family-owned chemical company, DuPont, was flush with profits amassed during World War I. One of DuPont's innovative products was a fast-drying automotive lacquer that was introduced in 1922. Up to this point the only paint that dried fast enough for mass production came in just one color, which gave rise to Henry Ford's famous statement that you could buy a Model T "in any color so long as it was black." In fact, early Model Ts were available in several other colors, but higher production volumes necessitated black paint because other colors required a longer drying time and hence massive dust-free storage facilities that would add considerably to production costs. DuPont's financial interest in General Motors gave the firm the inside track on the use of the new finishes. In 1924 GM cars began to appear in a variety of eye-catching hues, which highlighted how old-fashioned and dowdy the Model T had become.

As a final enhancement of its marketing efforts, General Motors took a leading role in the promotion of credit sales. While Ford had always insisted on cash payment for its cars, GM became the first automobile company to finance the sales of its products when it set up the General Motors Acceptance Corporation in 1919. GMAC allowed people to drive a new Buick or Oldsmobile now and pay for it later. Buying on time soon became common in the auto industry, as substantial numbers of cars were purchased for one-third down and twelve monthly installments. Refusing to submit to this trend, Ford stuck to its cash-only policy until 1928 in conjunction with the launch of the Model A. By this point, Ford was doomed to playing catch-up for many years to come. During the boom years of the 1920s, GM's organizational and marketing innovations propelled it to leadership of the American automobile industry, and by 1928 it held 47 percent of the American market.

## AND CHRYSLER MAKES THREE

General Motors was of course not the only manufacturer to compete with Ford. Chevrolet and Ford dominated the market for low-priced cars, but by paying a few hundred dollars more, American motorists could choose from a vast array of cars: Hupmobile, Kissel, Moon, Hudson, Essex, Studebaker, Nash, Willys-Overland, and Auburn, to name only a few. At the market's upper reaches, Ford took on GM's Cadillac division through its purchase of Lincoln in 1921. Buyers of luxury cars also could opt for a Packard, Peerless, Pierce-Arrow, or Franklin. But the real challenge to both Ford and General Motors came from the firm founded by Walter P. Chrysler. As was noted above, Chrysler had been a highly successful head of Buick until frustration with Durant's managerial style led him to resign from GM in 1920, and after a brief retirement followed by a sojourn at Willys-Overland, he went over to another firm struggling to survive, Maxwell. After becoming its president in 1923, he oversaw the design of the Chrysler Six, which appeared the following year. The Chrysler was built to high technical standards, employing a high-compression engine (its compression ratio was 4.7:1 instead of the 4:1 ratio commonly employed by other manufacturers) with a seven-bearing crankshaft assembly, four-wheel hydraulic brakes, balloon tires, and a replaceable oil filter.

Fortified by the sales success of the Six, Chrysler bought Dodge in 1928, which gave his firm expanded production capacity along with a well-regarded product in the medium-price range. Chrysler also aggressively went after the low-price segment through the introduction of a new brand, Plymouth. Fortuitously appearing in 1928 when Ford had nothing to offer potential customers while gearing up for the Model A, the Plymouth boasted a number of features that put it ahead of its low-priced rivals: four-wheel hydraulic brakes, aluminum pistons, and pressure lubrication. For the 1931 model year, it offered a significant innovation, a rubber-mounted engine that prevented vibration and torque reactions from being transmitted through the car and on to the driver and passengers. This feature allowed Plymouths to be marketed as having the "smoothness of an eight, the economy of a four." The Plymouth was an engineering and marketing success, and within a few years it accounted for more than half of the cars sold by the Chrysler Corporation.

## THE INDUSTRY BECOMES AN OLIGOPOLY

Chrysler's success ushered in a pattern that was to endure until the 1960s. A handful of small- to medium-size firms enjoyed sales and financial success through the 1920s, and a few of these were able to survive the lean years of the 1930s, but 80 percent of the market was now controlled by just three firms: General Motors, Chrysler, and Ford. In part this was due to the increasing technological sophistication of automobiles. In particular, the manufacture of enclosed metal bodies required a massive capital investment in hydraulic presses, metal-forming equipment, and electrical welding apparatus—expenditures that could not be borne by small firms. More than most industries, the automobile business came to exemplify the virtues of economies of scale. Heavy expenses were borne during the engineering of a vehicle and equipping a production line, but once this was done the marginal costs of assembling cars dropped steadily. This of course was not a new phenomenon; the Model T had dominated its market because its scale of production put it in a class by itself. Now the same thing was happening in the middle- and even upper-price bracket as General Motors and Chrysler took advantage of economies of scale to undercut their rivals in price and quality.

Also driving the industry toward oligopoly were the costs associated with effective marketing. In the early days of the industry, manufacturers could sell their products with little or no advertising. A famous slogan used by the Packard Motor Car Company originated with the firm's terse response to a potential customer's request for a descriptive brochure: "Ask the man who owns one." This might have been sufficient during the early years of the automobile industry, but the 1920s marked the beginning of the media age, where radio and mass-circulation magazines had become essential components of effective marketing, and as such commanded significant financial outlays. The market for automobiles had become national and even international in scope; a firm could not survive by catering to a regional market that remained in close proximity to its production facilities, hence the need for advertisements that reached from coast to coast. Finally, a manufacturer could reach more potential customers when it had a large network of franchised dealers, and this in turn required high-volume production to stock dealers' lots. The seller's market of the twentieth century's first three decades allowed a fair number of small- and medium-sized firms to survive and even prosper, but their fate was sealed with the arrival of hard financial times in the 1930s.

## EUROPE VERSUS AMERICA

Market concentration was less evident in Europe during the 1920s, although it was clearly advancing. In Britain, for example, six firms (Austin,

Ford, Morris, Rootes, Standard, and Vauxhall) accounted for about 80 percent of the home market in 1930. Economies of scale were less of an issue because the market for automobiles was much smaller than it was in the United States. Although Europe's population was roughly equivalent to that of the United States, weaker purchasing power limited the number of potential customers. In 1927, a year that the United States had one car for every 5.3 persons, the ratio in England and France was one to forty-four (Bardou et al. 1982, 112).

Many of the cars that plied European roads in the 1920s were American designs. Some of them had been exported from the United States; although America's car industry primarily served a domestic market, foreign sales were not trivial, accounting for more than 11 percent of the industry's output in 1929. Faced with a deluge of cheap, mass-produced American cars, many European governments attempted to limit imports through the imposition of high tariffs, such as the 33⅓ duty that Britain levied on imported cars. This was not a uniquely European situation; the United States had long imposed a 45 percent tariff on imported automobiles (reduced to 30 percent in 1913 for cars costing under $2,000). Given the productive efficiency of American firms and the cost advantages it conferred, it is not likely that the free importation of European vehicles would have made much of a difference in the U.S. market, but European tariffs were a definite obstacle for American exports. In order to circumvent these restrictions, American firms set up overseas factories for the manufacture of their cars. Ford was one of the first to do so when in 1911 it set up an English facility for the assembly of Model Ts in Manchester. In the years that followed, Ford's reach became truly global, and by 1929 it had assembly plants in Canada, Ireland, Belgium, Denmark, France, Italy, Argentina, Brazil, Chile, South Africa, Australia, Malaya (now part of Malaysia), Mexico, Uruguay, India, Turkey, Germany, Spain, and Japan.

Although the Model T had been designed specifically for American conditions, it was highly successful abroad. In Britain, for example, it outsold all other cars until 1923. General Motors pursued a different strategy by buying two established European firms, Britain's Vauxhall in 1925 and Germany's Opel in 1929. Both Ford and GM also set up factories in Japan (in 1925 and 1927, respectively), and in so doing became that nation's largest source of locally produced automobiles.

European firms did not stand by idly as American firms grabbed a substantial share of their home markets. Although the Model T was Europe's best-selling car by a large margin, it was better suited to bouncing down rural roads than navigating the narrow lanes of European cities. Smaller cars ultimately proved to be more appropriate for European conditions, especially when high gasoline taxes penalized American cars that had not been designed with fuel economy in mind.

Making matters more difficult for the sale of American cars, the British government introduced a "horsepower" tax in 1921. The tax was not based on an actual measurement of an engine's power output; rather, it was based on a formula previously developed by the Royal Automobile Club. According to this formula, horsepower was defined as $D^2N/2.5$, where D was the engine's bore, N was the number of cylinders, and 2.5 was a more or less arbitrary figure that was thought to be "reasonable and sufficiently accurate for comparative purposes" (quoted in Wood 1988, 43). By assessing a tax of £1 per "horsepower," the government put American-style cars with their large-displacement engines at a significant cost disadvantage. But in the long run, this tax also harmed British manufacturers. Since the formula took into account only an engine's bore and number of cylinders, British cars were characteristically powered by small-displacement engines with a small bore but a long stroke. Engines of this sort were adequate for British conditions where highway speeds tended to be low, but they were not well suited to the higher speeds or severe driving conditions often found in other parts of the world.

The combination of high taxes and more constricted driving conditions resulted in a significant divergence of American and European car designs, one that has held to the current day. Some European manufacturers like Rolls-Royce and Hispano-Suiza made large, luxurious cars, but the typical European car was much smaller than its American counterpart. What has often been heralded as the first successful European small car was the Bébé Peugeot. Designed by Ettore Bugatti and introduced in 1912, it was powered by a tiny 855-cc, four-cylinder engine. It was a great sales success, allowing Peugeot to overtake Renault as Europe's largest car manufacturer during the following year. In England, Herbert Austin produced what has been described as Europe's first mass-produced car, while his firm's great rival, Morris Motors, was able to overtake Ford as Britain's largest car manufacturer on the strength of its diminutive Morris Cowley. These were significant accomplishments, but they should not obscure the fact that British and Continental manufacturers were not playing in the same league as the large American firms. Fiat, which accounted for nearly three-quarters of car sales in Italy in the late 1920s, produced only 54,000 cars in 1929 (Bardou et al. 1982, 108), while Morris held 41 percent of the British market 1925 when it produced 54,151 automobiles (Wood 1988, 38), rather unimpressive figures when put alongside the 1,643,295 cars that Ford produced in 1925.

Although European firms did not lack for technical sophistication in the design of their cars, they seriously lagged behind the United States when it came to efficient production. In 1927, for example, French manufacturers needed 300 man-days of labor to assemble a car, whereas U.S. firms needed only seventy (Bardou et al. 1982, 102). Notably lacking in Europe was the signature technology of the American automobile industry, the assembly line. Citroën, the manufacturer most dedicated to replicating American production technologies, began using an assembly line in 1919, but other firms were more hesitant. Opel, Germany's largest manufacturer, put in an assembly line in 1924, and Austin and Morris waited until the 1930s before doing so.

The small production volumes of European manufacturers reflected the vicious circle in which they were caught. A relatively small market prevented the exploitation of economies of scale, but the lack of these economies resulted in high prices and restricted sales. As we shall see, it was not until the 1950s and 1960s that the circle was broken and Europe became an automotive mass market.

## TECHNOLOGICAL ADVANCES

Although technological progress may not have been as rapid as it had been during the century's first decade and a half, automotive engineering still moved ahead at an impressive clip. For the average motorist, the biggest change was the closed automobile body. In 1920, 90 percent of cars produced were roadsters, touring cars, and other topless automobiles. Some protection could be afforded by fabric tops and side curtains, but these were awkward to deploy, and they did a highly imperfect job of keeping out dust, wind, and rain. Closed cars had been around for much of the automobile's history, but most of them were overweight, awkward looking, and priced out of the reach of the average motorist. Advances in the production of sheet steel and plate glass brought down costs, and in 1921 Essex introduced the first popularly priced enclosed car. Its low price, only $300 more than the touring car version, was made possible by using straight body panels and simplified construction that largely eliminated compound curves for the body's wooden framing.

In 1925 more closed cars were sold than open ones, and by 1929 the ratio of closed to open cars had been reversed; 90 percent of the cars sold were enclosed models. As enclosed automobiles became the standard, cars served more effectively as everyday transportation because they could be driven in all sorts of weather without requiring the bundling up of drivers and passengers. In addition, automobile usage expanded as fewer motorists felt the need to put their cars in storage for the winter, a common practice in many parts of the country.

The switch to enclosed cars was paralleled by changes in the way car bodies were made. Although wooden-framed bodies continued to be used throughout the 1920s, Dodge had already introduced the all-steel body, which was developed with the assistance of Edward G. Budd (1870–1945), a Philadelphia manufacturer. First used in 1914, it was stronger and less prone to developing disconcerting creaks and rattles. It was also cheaper to build because it could be painted with baked enamel paints that eliminated a great amount of hand finishing. Dodge began to manufacture enclosed metal bodies in 1923, although the roof consisted of fabric stretched over a wooden framework.

In addition to offering practical advantages, enclosed cars also offered automobile stylists the opportunity to create more integrated body designs. Open cars could be quite pleasing aesthetically, but the effect often was ruined as soon as the top was put up. Cars styled in the 1920s were still based on the idea that the major body elements—engine compartment, fenders, trunk, and passenger compartment—should retain their separate identity, but within this idiom there was an effort to produce more visually appearing bodies. The main stylistic trend during this period was to soften the angularity of car designs, and, as noted above, to create both the illusion and the reality of longer and lower automobiles.

Contributing to the lowering of cars was the introduction of balloon tires in the early 1920s. These tires, which were mounted on smaller diameter wheels, had wider treads and sidewalls, and ran at inflation pressures of 30 lbs. per square inch, less than half the pressure of the tires they replaced. The result was a longer lasting tire that also provided a smoother ride, albeit with some loss of steering responsiveness. It was also during the 1920s that metal disc wheels began to replace wooden-spoke artillery wheels. Wire wheels also became more popular after being largely forgotten during the century's first decade.

In regard to the automobile's basic running gear, the mid-1920s are notable for the widespread adoption of front-wheel brakes. These had been tried out during the previous decade, but a number of problems, most notably an occasional tendency for the brakes to apply themselves when the front wheels had been turned to full lock, had given them a bad reputation. Hydraulic brakes, another key element of an effective braking system, made their appearance in the early 1920s. Invented by Malcolm Loughead for use on aircraft, they were effectively used by a Duesenberg that won the French Grand Prix in 1921. By 1927 virtually every car sold in America had four-wheel brakes as standard equipment. Ettore Bugatti may have dismissed complaints about the poor braking performance of some of his cars by blustering that he "made cars to go, not to stop," but there can be no doubt that better brakes allowed cars to travel faster and with greater safety.

Many of the performance gains of the engines of this era can be attributed to the fuel that they used. It had long been understood that the power and efficiency of an engine could be improved by increasing its compression ratio. However, there were limits to how high this could go; too much compression resulted in an engine that "knocked" or "pinged," and in extreme cases caused serious damage to the pistons. Increasing the octane rating of gasoline could alleviate this problem, but this resulted in more expensive fuel. In 1919 Charles F. Kettering and Thomas Midgley Jr. of the General Motors Research Laboratory began to study the causes of knock in the hope of finding a way to prevent it. Their efforts resulted in the discovery that the addition of tetraethyl lead to gasoline significantly increased knock resistance. Gasoline with tetraethyl lead was first sold commercially in early 1923, and over the years it was one of the chief reasons that engines developed more power and operated more efficiently. As we shall see, however, tetraethyl lead was very much a mixed blessing, and its use eventually had to be banned.

## THE AUTOMOBILE AND RECREATION

From its earliest days, much of the automobile's appeal has rested with the opportunities it has afforded for recreation, adventure, and fun. As automobile ownership rapidly expanded, motorized travel became a common feature of American life. Automobile touring in this era of open cars required hardiness and a sense of adventure, and as such it formed a natural complement to camping out. But even automobile tourists who preferred more luxurious accommodations were likely to be thwarted because hotels were not eager to receive disheveled motorists, and even if they did, they rarely offered adequate parking facilities. Faced with these obstacles, travelers often pitched their tents at any convenient roadside location, a behavior that often angered property owners. To obviate this problem, many towns set up campgrounds in the hope of attracting travelers who were likely to patronize local businesses during their stay. By the early 1920s the United States had more than 1,000 of these campsites, but local residents began to get second thoughts when perpetual transients, migratory workers, and even criminals started showing up. In order to discourage these "undesirables," local governments began to set time limits and charge fees for campground use. This created an opening for the private sector, and before long local entrepreneurs were setting up commercial camping facilities for motorists.

Even with a predictable supply of campgrounds, auto camping was a bit too much for many travelers, requiring the packing of cooking gear, tents, and other necessities, and the pitching of the tent and the preparation of a meal at the end of a long day's travel. It therefore did not take much of a conceptual leap for some campground owners to realize that there was an unsatisfied need for something better than a plot of land upon which to put up a tent. By the mid-1920s, campground operators were offering permanent shelters for their clients. Many of these were little more than hastily erected shacks or even converted chicken coops, but before long these spartan facilities had evolved into more luxurious accommodations complete with innerspring mattresses, kitchenettes, refrigerators, and even attached garages. These facilities often were known as "tourist cabins," but in 1925 the English language was enriched with a new word when the Motel Inn opened in San Louis Obispo, California. This was really a conventional hotel that incorporated space for parking, but in the years that followed the word "motel" appeared on thousands of neon signs as lodgings specifically designed for automobile travelers took root alongside the nation's highways.

Paralleling the spread of motels was the proliferation of many other roadside establishments. Although serving food to people in their vehicles goes back to horse-and-buggy days, credit for being the first drive-in restaurant is usually given to J. G. Kirby's Pig Stand restaurant that opened along the Dallas–Fort Worth highway in 1921. As drive-ins and other commercial establishments catering to motorists became increasingly prominent, an amazing variety of vernacular architectures began to spring up. Drivers and passengers traveling at highway speeds had only a few seconds to view a roadside establishment and then decide to patronize it. Large signs were one way of attracting attention, but even better were structures that caught the eyes of automobile travelers: hamburger stands in the shape of dogs, permanently grounded stucco airplanes, and motel rooms that mimicked wigwams. The automobile was reshaping the landscape, not always in the most tasteful and aesthetically pleasing way.

## COMMUNITIES TRANSFORMED

The great expansion of automobile ownership that continued through the immediate post–World War I era could not fail to have important consequences for American society. In the 1920s the sociologists Helen and Robert Lynd conducted a study of Muncie, Indiana, which they dubbed "Middletown." Their research covered many facets of life in the Midwestern town and how it was changing. One source of social and cultural change stood out; as one resident declared to the Lynds, "Why on earth do you need to study what's changing this country? I can tell you what's happening in just four letters: A-U-T-O!" (quoted in Lynd and Lynd 1929, 251).

The Lynds' study described many consequences of automobile ownership, such as an apparent decline in neighborly interaction and the financial sacrifices people made in order to pay for their cars. Of special concern were the effects of the automobile on conventional morality, which led one judge of the juvenile court to refer to the automobile as "a house of prostitution on wheels" (quoted in Lynd and Lynd 1929, 114). In fact, in Middletown and elsewhere, most amorous uses of the automobile did not involve sex for payment. The significance of widespread automobile ownership for sexual behavior was that it allowed young men and women to escape close supervision. It is impossible to ascertain the extent to which premarital sex increased in the 1920s, and how much of it was due to the spread of car ownership. It seems beyond question, however, that the availability of cars made a strong contribution to the changes in behavior and moral standards that characterized the era.

Another important cultural shift during that era was the dissolution of traditional ideas regarding the proper place of women in society. The wartime contributions of women had demonstrated the fallacy of their being viewed as "the weaker sex," and the Armistice of 1918 did not result in a return to the attitudes that prevailed before the war. Many women wanted more from life than what the traditional roles of wife and mother offered, and driving a car could be an important way of asserting one's capabilities and the autonomy that went with it. Motivated in part by a desire for greater freedom and self-reliance, women settled into the driver's seat in numbers that would have seemed inconceivable before the war.

The automobile thus served as a key artifact in what today would be called women's liberation. At the same time, however, driving a car did not guarantee an escape from traditional sex roles; in fact, it could just as easily reinforce them. For every adventurous flapper at the wheel of a smart new roadster, there were many other women for whom the car was primarily an instrument for shopping, conveying children, and engaging in traditional household tasks. One national survey found that urban women were spending about seven and one-half hours a week in household-related driving (Scharff 1991, 151). Important as the role it played in changing culture and social behavior, automobile use also reflected well-established patterns that were slow to change.

By the mid-1920s, the manufacture of automobiles had become America's largest industry. In supplying what seemed to be an endless demand for private cars, automobile manufacture accounted for 80 percent of the rubber industry's output, 75 percent of glass, 25 percent of machine tool purchases, and 20 percent of steel, while the more than 17 million cars plying America's roads consumed 90 percent of the nation's gasoline output. The importance of the car for the American economy and society was underscored in 1929 when the U.S. automobile industry set a new record by producing 5,337,087 cars. More than with any other consumer good, widespread automobile ownership exemplified the unparalleled prosperity of a large segment of American society. To be sure, tens of millions still lived in poverty, but as the 1920s drew to a close, the prevailing view was one of endless improvement in the American standard of living, with "a chicken in every pot, and two cars in every garage." Yet as the 1920s were coming to a close, it all came crashing down. Two decades were to elapse before the industry equaled the production record that it had set in 1929. What lay ahead was economic depression and war, and with them came far-reaching changes to automobiles and the industry that produced them.

4

# Hard Years and Heroic Days, 1929–1945

◆

As the 1920s drew to a close, there seemed to be every reason to believe that prosperity would continue unabated. National income was rising, new industries like radio and commercial aviation were emerging, and the stock market was soaring to heights never seen before. Then it all began to unravel on October 29, 1929, when the collapse of the American stock market heralded a decade marked by economic collapse, massive unemployment, and widespread social disruption. Paradoxically, however, the severe economic downturn did not put an end to automotive progress. While automobile sales declined sharply, the pace of engineering and stylistic innovation did not slacken. In Europe the economic situation was somewhat better, and although automobile ownership still did not reach American levels, it was coming within the reach of greater numbers of people. The hard economic times of the Great Depression also were the scene of the rise of organized labor, and on a less serious note, automotive-based novelties like trailers and drive-in movies came to the fore. By the end of the decade the worst days of the Depression had passed, only to be replaced by the most extensive and destructive war in human history. In ultimately winning that war, the Allied powers leaned heavily on their automobile industries, which more than met the challenges that had been put before them.

## THE BOOM GOES BUST

The American automobile industry faithfully mirrored the severe contraction of the economy as a whole, as indicated by the dismal record of sales through the decade as Table 4.1 illustrates. As can be seen in these figures, the rebound that began in the middle of the decade proved to be a false dawn, as a severe recession that began in 1938 wiped out most of the gains of the years leading up to it. Overall registration of automobiles also declined from 1930 to 1934 because many people found that they could no longer afford to operate and maintain their cars. Total registrations returned to their 1929 level in 1936, and by 1939 they had gone up by slightly more than 3 million, an increase of about 11 percent over 1929. Significantly, motorists did not curtail their driving, however; the percentage increase in gasoline consumption was more than twice as much as the increase in the number of cars on American roads during this period.

While the impact of hard economic times severely affected car sales, the cause-and-effect equation also went the other way. In the mid-1920s some automobile industry executives had begun to ruminate on the imminent saturation of the market for their products, and by the end of the decade it was evident that the industry had more production capacity than the market could continue to absorb. Massive capital expenditures to expand production had helped to fuel the economic boom of the 1920s, but these could not be sustained in an era when the sales of cars were bound to level off. First-time buyers were now a minority as most car purchases were replacement sales, and many of these were for used cars, not new models. Making a bad situation worse was an increasingly unequal distribution of income that prevented a large segment of the population from buying new

**Table 4.1**
**Total Automobile Sales (1929–1939)**

| | |
|---|---|
| 1929 | 4,445,178 |
| 1930 | 2,787,456 |
| 1931 | 1,948,164 |
| 1932 | 1,103,557 |
| 1933 | 1,560,599 |
| 1934 | 2,160,855 |
| 1935 | 3,273,874 |
| 1936 | 3,679,242 |
| 1937 | 3,929,203 |
| 1938 | 2,019,566 |
| 1939 | 2,888,512 |

cars even if they wanted to. Buying on credit made the purchase of a new car possible for those who were a bit better off. This was not a problem when people could make their monthly payments, but when they lost their jobs or were otherwise financially pinched, defaulted loans mushroomed and sales dropped further. In sum, economic contraction and declining car sales fed on each other, significantly reinforcing the other causes of the Depression, and making it deeper and longer than it otherwise might have been.

## AN INDUSTRY CONTRACTS

For automobile aficionados, the 1930s are most often remembered for the loss of many long-established automobile manufacturers; gone from the luxury field were Peerless (1932), Pierce-Arrow (1938), Duesenberg (1937), Franklin (1934), Marmon (1933), and Stutz (1938). One noteworthy new design was the creation of Erret Lobban Cord. The cars that bore his name were technically unorthodox, using front-wheel drive and, in some cases, supercharged engines. Although eagerly sought by collectors today, they failed to appeal to contemporary car buyers, and production ceased in 1937. Medium-priced cars fared no better. Auburn, another part of E. L. Cord's empire, was known for the high quality and outstanding value of its products, but it ceased production in 1936. Reo, a firm started by Ransom E. Olds (hence the name) after a 1904 disagreement with his partners at Oldsmobile, gave up on cars in order to concentrate on trucks in 1936. Graham-Paige introduced the first moderately priced supercharged car in 1934, and went on to make more supercharged models than any other manufacturer; it barely survived the 1930s, ceasing production in 1940. The same fate befell Hupmobile, which also called it quits in 1940. At the lowest end of the scale was the American Austin. Based on one of Britain's most popular cars, the Austin 7, it was tiny by American standards, weighing only 1,200 pounds. Purchased from bankruptcy court in 1935 and renamed American Bantam, its reprieve was brief, lasting only to 1941. However, as we shall see, its manufacturer went on to create one of the great automotive designs of all time.

Other makers soldiered on. Willys, once the third-largest American automobile manufacturer, declared bankruptcy in 1933 and then staggered through the rest of the decade after being reorganized under new management. Packard was able to survive the 1930s by marketing a less expensive line of cars to go with its luxury models, to the long-term detriment of its image as a manufacturer of uncompromisingly high-quality automobiles. Hudson was kept going by the success of its Terraplane, a low-priced car with excellent performance, but the strain of keeping the firm alive probably contributed to the premature death of its president, Roy Chapin. Nash recovered some of the sales lost to the Depression through the introduction of a low-priced line called LaFayette. It also got into the kitchen appliance business through its 1937 acquisition of Kelvinator, although it was a money-loser at first. (General Motors had done the same thing by acquiring Frigidaire in 1919 on William Durant's conviction that both cars and refrigerators were essentially the same thing: metal boxes with motors inside.) Studebaker went into receivership in 1933 and its president subsequently committed suicide, but it subsequently was able to stay afloat for more than two decades. During the 1930s the firm had the dubious distinction of using a singularly ill-chosen name to designate one of its models: the Dictator.

The shakeout of the 1930s reinforced a trend that had become evident during the previous decade. The American automobile industry was a classic oligopoly where the "Big Three" of General Motors, Chrysler, and Ford now accounted for about 85 percent of all automobile sales, with General Motors alone holding about 40 percent of the total. The "independents" (primarily Nash, Studebaker, Hudson, and Packard) were left to fight over what was left.

## THE AUTOMOBILE BUSINESS IN EUROPE AND ASIA

The European market did not undergo as catastrophic a tumble during the 1930s; in Britain and Germany annual car sales actually increased from 1929 to 1939. Despite advancing sales, Europe still had not entered the era of mass motorization. Although car ownership became more widespread, the gap between the car-population ratio in the United States and Europe improved only to the point where it had been in 1913. Most people still got around on trolleys, buses, and bicycles, or motorcycles if they were a bit more affluent.

An expanding market did not benefit all manufacturers. As was happening in the United States, many firms fell by the wayside as oligopoly prevailed. The Big Six (Austin, Ford, Morris, Rootes, Standard, and Vauxhall) accounted for 90 percent of the British market in 1939, while in France, Renault, Citroën, and Peugeot held 73 percent of their national market (Bardou et al. 1982, 146). In Italy a single firm, Fiat, controlled nearly 85 percent of the domestic market (Bardou et al. 1982, 147).

Concentration was not confined to the industry's constituent firms; in a few countries, automobile manufacture disappeared altogether. Most notably, the Depression put an end to the automobile industries of Belgium and Switzerland, whose luxury products were particularly vulnerable. At the same time, however, the 1930s saw significant growth in the industries of two nations that were latecomers to large-scale automobile manufacture, the Soviet Union and Japan. Russia had been an economic and technological backwater prior to the Bolshevik Revolution of 1917. Only a few thousand automobiles existed in a vast country that stretched from the Black Sea to the Pacific, the majority of them imported or built in factories set up by foreign firms. The first years of Soviet government were devoted to establishing a basic industrial infrastructure; truck, bus, and automobile production did not begin in earnest until the late 1920s. Ironically, American machine tools and production expertise provided the foundation for the Soviet automobile industry, one important example being the key role played by Ford in the construction of a plant in Moscow in 1930–1931 with an annual capacity of 24,000 cars. By 1939, the Soviet automobile industry was able to produce more than 200,000 cars a year, although the main emphasis was on truck production (Bardou et al. 1982, 148ff).

Another rising power, Japan, also was in the initial stages of building an automobile industry. Although indigenous attempts at automobile production dated back to the first decade of the century, by 1936 assembly plants established by Ford and General Motors accounted for 90 percent of the market (Flink 1988, 271). This situation was not acceptable to an increasingly nationalistic and militaristic government, which began to take strong measures to freeze out the American firms and encourage domestic manufacturers. The most important of these were Toyota and Nissan. Toyota, which began producing significant numbers of cars in 1937, was an offshoot of the Toyoda Automatic Loom Works. The firm that eventually was known as Nissan began producing a small car called the DAT in 1914. In 1931 a car known as the Datson (son of DAT) appeared; this name was converted to "Datsun" a year later and was given to a car closely resembling the English Austin 7, which had been copied without the firm's permission. Toyota and Nissan produced only a few hundred cars, and toward the end of the decade they had shifted almost exclusively to the production of trucks. Neither firm was in a position to determine what to produce; the military-dominated government dictated industrial policy, and it wanted trucks to supply the armed forces that took control of Manchuria in 1931 and invaded the rest of China in 1937. Efforts to manufacture trucks were not completely successful. Beset by raw material and labor shortages, Japanese industry failed to provide all of the vehicles that were needed. This shortage was a major cause of the military reverses suffered by the Japanese during

World War II, as what initially seemed to be an unstoppable force found itself severely hindered by chronic shortages of essential supplies.

## ENGINEERING AND DESIGN

Although cars of the 1930s generally performed better than those of the previous decade, the Depression era was not marked by epochal engineering breakthroughs. Although the typical car of 1935 was faster and more comfortable than its 1925 counterpart, most of the difference was due to steady development and the application of existing technologies rather than radical innovations. This can be seen in the evolution of braking systems. Brakes using hydraulic instead of mechanical actuation dated back to the early 1920s, but in 1930 only nine out of thirty American manufacturers used them. General Motors made hydraulic brakes standard equipment in 1934, while Ford waited until 1939 before adopting them, and some European firms were even slower to do so. In similar fashion, independent front suspensions became more common toward the end of the 1930s, but many major manufacturers, most notably Ford, retained their solid-axle setups.

If basic mechanical features did not change dramatically during the 1930s, the same cannot be said of the outward appearance of automobiles. The typical car of the early 1930s was tall, upright, and generally rectangular. Many of its key elements—headlights, fenders, bumpers, running boards, radiator, and trunk—were separate entities that literally stood out from the car's body. By the end of the decade, the appearance of cars had changed dramatically; their basic shape was lower and more integrated, with sweeping lines and a rounder, streamlined appearance.

Enthusiasm for streamlining was motivated in part by aviation technology. During the late 1920s and early 1930s, aircraft rapidly evolved from open-cockpit, wood-and-fabric biplanes to low-wing, all-metal monoplanes with wind-cheating cowled engines and retractable landing gear. The benefits were evident, as streamlined airplanes like the Douglas DC-3 were able to fly faster and higher than their predecessors. But more than simply promising higher performance through efficient design, streamlining offered an escape from the Depression. According to one line of thought, the Depression was the result of obsolete social institutions that blocked social and economic progress in the same way that air resistance impeded the motion of unstreamlined aircraft. Over and over, one could hear appeals to "streamline" institutions, rules, and social arrangements of every kind. Streamlining thus promised at least a metaphorical relief from the frictions and obstacles that were preventing an escape from the Depression.

By the second half of the 1930s, the ideology of streamlining was being realized in airplanes, ships, locomotives, buildings, furniture, and even household appliances. Although a few automobiles had been designed to minimize air resistance at an early date, they were oddities, and it wasn't until the mid-1930s that streamlining became a major influence on the design of car bodies. At first, streamlining meant rounded-off corners, bulbous front fenders, and windshields and radiators that sloped back slightly. Then in 1934 the Chrysler Corporation took a bold step with the introduction of the radical Airflow. Marketed with minor variations as a De Soto and a Chrysler, the Airflow represented a major break from conventional design. Consciously designed to minimize air resistance, the Airflow looked like no other car on the road. Unfortunately for Chrysler, the buying public found it too different. Radical designs have often met with customer resistance, and this was compounded in the case of the Airflow by the inescapable fact that by any aesthetic standard the car was ugly. Sales were disappointing in the extreme, and 1937 marked the last year of the car's production.

One of the main reasons for the rejection of the Airflow was that the front of the car sat ahead of the wheels, the reverse of the design idiom that had held sway for many years. This had been done to create more room for

Although it embodied a number of advanced design and engineering features, the 1934 Chrysler Airflow was a commercial failure. © Bettman/CORBIS.

the rear-seat passengers by moving the passenger area forward relative to the wheelbase. It also improved weight distribution to 54 percent on the front wheels and 46 percent on the rear wheels, reversing the typical tail-heavy weight distribution of the period, an important advantage at a time when highway speeds were rising and balloon tires had altered handling characteristics. But it also resulted in a nose-heavy look that turned off potential customers. Later designs avoided this stylistic problem by using the radiator grille as a key design element. Since it had no functional purpose, the grille could take on a variety of shapes and forms. Stylists took advantage of this plasticity in a number of ways; some grilles had vertical bars, while on others they were horizontal. Some grilles swept around the front of the car; others looked like fencers' masks. Whatever their exact form, they visually lightened the front while at the same time made it easier to distinguish one make of car from another.

The marketing failure of the Airflow underscored the importance of making only incremental design changes. Successful automobile designers understood that customers shied away from designs that marked too radical a departure from the norm. Consequently, the phrase "most advanced yet acceptable" (MAYA) governed the design process.

The evolutionary development of 1930s car design can be seen in the fate of the front fenders. In 1930 fenders were freestanding components that closely followed the contours of the wheels they covered. As the decade progressed they became important stylistic elements, elongated forms that sometimes swept all the way back toward the middle of a car's body. In some cases, the streamlined appearance of fenders was augmented by fitting them with skirts that covered a large portion of the wheel well. The evolutionary process of automobile styling can also be seen in the gradual reduction in the size of running boards, which in any event were becoming superfluous as car bodies moved lower to the ground.

Although streamlining may have stimulated sales, it had some negative consequences. Visibility suffered as roofs were lowered, windshields were tilted at a sharper angle, and pillars were made thicker, all of which conspired to reduce glass area. Curved front ends necessitated the use of "alligator" hoods over the engine compartment instead of the typical prior arrangement, two hatches that were horizontally hinged at the top. As a result, side access was restricted and mechanics had to resign themselves to a greater frequency of skinned knuckles when working within the lower reaches of an engine compartment.

The basic construction technology of car bodies also began to change as a number of manufacturers took up unit construction. This entailed abandoning body-on-frame construction and designing the body of a car so that the entire structure provided the attachment points for the engine, suspension, and other major components. This resulted in more strength and rigidity. At the same time, however, unitized bodies made for rigidity in another sense of the word because the costs entailed in setting up a production line for unit construction were considerably greater than those required for traditional frame-and-body construction. This meant that a car had to be produced in large volumes to justify the additional manufacturing expense. Production runs had to be longer, the variety of models fewer, and a poorly selling car could result in a substantial financial drain.

Credit for being the first unit-construction car usually goes to Italy's Lancia Lambda, which appeared in 1922. The Lancia was produced in small quantities, but since it was an expensive car, the additional cost of production could be folded into the selling price. Vincenzo Lancia received a patent for unit construction in 1919, and the description in his application provides a good summary of its essential principles: "A type of car in which the chassis frame is eliminated, the connection between the back axle and front axle comprising a rigid shell which fulfills the functions of a body in a normal car" (quoted in Tubbs n.d., 3). The first unit-construction car to be manufactured in large quantities was the 1935 Opel Olympia produced by General Motors' German division, although the aforementioned Chrysler Airflow made use of what might be described as semi-unit construction because the car's suspension was mounted to a separate subframe.

An outstanding example of a car using unit construction was the Citroën Traction Avant. Introduced in 1934, it was one of the first cars to successfully employ front-wheel drive, which gave it excellent handling and roadholding. Although it was an engineering triumph, the cost of engineering and tooling up for the car was beyond the firm's financial capability, and the car's introduction was soon followed by bankruptcy and the premature death of Andre Citroën, the firm's founder.

The Citroën shared with other unitized cars one problem that could cause considerable grief to its owners, a loss of structural integrity due to rust; if the "tinworm" got a substantial toehold, the car could disintegrate rapidly. Unit-construction cars were also more difficult to repair if they were involved in an accident. If a body shop lacked the necessary equipment and expertise, even a relatively minor collision could result in a car that was permanently out of kilter.

## THE FORD V-8

The massive dislocation and retooling that preceded the production of Ford's Model A had not been in vain. In 1928, its first year of production, 820,000 Model As were made, and Ford went on to outsell rival Chevrolet for the next two years. In all, 5 million Model As were built. But unlike the Model T, which remained in production for nineteen years, the Model A lasted only through the 1931 model year. Henry Ford had been working on a radical X-8 engine (four banks of two cylinders arranged in the shape of an "X") that he hoped would be the basis of a new line of cars. When it proved to be unworkable, he turned his attention to a more conventional configuration, a V-8 (two banks of four cylinders arranged in the shape of a "V"). Engines of this configuration had been produced in limited numbers since the early twentieth century, but they had been restricted to high-priced automobiles; Ford was the first to put a V-8 engine in a low-priced car. The V-8 that Ford's engineers designed represented an advance in production technology. Just as the Model T's engine had been one of the first to have a cylinder block cast as a single unit, the new Ford engine was one of the first mass-produced V-8 engines to have a block cast in one piece. Producing the engine proved to be a formidable task. Making the engine block entailed the precise positioning of fifty-four sand cores for each casting, and keeping these cores in place during casting was a major headache.

Henry Ford and the famous flathead Ford V-8. His son Edsel's attention is directed elsewhere. © Underwood & Underwood/CORBIS.

After a great deal of effort directed by Charles Sorenson, Ford's chief engineer, the first cars equipped with the new engine began to move off the assembly line in March 1932. Although the V-8 used side valves instead of more efficient overhead valves (for this reason it is often referred to as a "flathead"), its conservatively rated 65 hp put it well ahead of the competition. As one of its fans, the bank robber Clyde Barrow, gushed in a letter to Ford, "For sustained speed and freedom from trouble the Ford has got ever [sic] car skinned and even if my business hasn't been strictly legal it don't hurt anything [sic] to tell you what a fine car you got in the V-8" (Flink 1988, 230–31). The basic design stayed in production until 1953 and was the starting point for thousands of hotrodders who were able to extract far more power from these engines than had ever been intended by their manufacturer.

## EASIER SHIFTING

As with most other cars of the time, the V-8 engine was hooked up to a friction-plate clutch and a three-speed manual transmission. This was not a very satisfactory arrangement for many drivers, for whom the most difficult thing about driving a car was working the clutch and manually shifting from one gear to another. Many drivers never really mastered the technique, so automobile manufacturers endowed their products with big engines that produced abundant torque as a way of reducing the need to shift gears. Some firms even publicized this capability through long-distance trips that their cars made while using only high gear. This approach was most often employed by American cars because cheap gasoline and the absence of a horsepower or displacement tax allowed the use of large, thirsty engines. It was not an option in most European countries, and drivers had to resign themselves to frequent use of the gear lever.

Some of the pain of using a manual gearbox was alleviated by the introduction of synchromesh gearboxes. Before the introduction of synchromesh, drivers had to carefully match the speed of the engine with the rotational speed of the gears being selected by using the technique of double clutching. With synchromesh all that was necessary was to depress the clutch, move the gear lever, and release the clutch. First gear usually was unsynchronized since it was used primarily for accelerating from a standstill. Synchromesh was introduced on the 1929 Cadillac and LaSalle, and was a universal feature of American cars by 1932. It diffused more slowly in European makes of automobiles, some of which waited until the post–World War II era before providing their cars with this feature.

For many motorists, synchromesh simply made the best of a bad situation; what they really wanted was a transmission that shifted itself. As with many automobile innovations, automatic transmissions date back to the early days of the industry, but many years were to elapse before they became practical propositions. In Europe, preselector gearboxes were employed in some high-priced cars, while semiautomatic transmissions were used by some American cars, but these still required the use of a clutch pedal. Credit for building the first effective fully automatic transmissions goes to General Motors, which after a decade of development introduced the Hydra-Matic on the 1939 Oldsmobile. A complex design that used a fluid coupling along with a set of clutches and bands, it exemplified how what we now call "user friendliness" often requires the employment of much more complicated mechanisms. The Hydra-Matic ended up being used by Cadillac and Pontiac, as well as a number of non-GM firms (Ford's Lincoln Division among them) that were saved the expense of developing their own automatic transmissions for a number of years.

## THE DIESEL ENGINE

The original impetus for the diesel engine was Rudolf Diesel's desire to design an optimally efficient engine through the application of the principles of thermodynamics. Diesel never realized this goal, but after a great deal of effort he succeeded in designing a type of engine that operated with greater thermodynamic efficiency than the Otto-cycle engine. In a diesel engine, the air-fuel mixture is ignited by the heat that is produced as air is compressed in the combustion chamber, and diesel engines operate more efficiently because they use higher compression ratios than conventional spark-ignited engines. Diesels were primarily used as stationary engines and to power ships and locomotives during the first three decades of the century. Benz produced a diesel-powered truck in 1923, and Daimler-Benz produced a diesel-powered car, the Mercedes 260D, in 1936. Citroën followed with a diesel-powered car the following year. By the late 1930s Mercedes had built 2,000 diesel-powered cars, most of them destined for use as taxis. For many years, however, the high torque and low fuel consumption of diesels made them most suited for use as truck engines. In 1931 Clessie Cummins (1888–1968), the founder of Cummins Diesel, demonstrated the diesel's economy and practicality by driving a diesel-powered truck from New York to Los Angeles, a distance of 3,214 miles (5,172 km), in 97½ hours at a fuel cost of only $11.22. As a further demonstration of the diesel's capabilities, he sponsored diesel-powered race cars that participated in the Indianapolis 500 in 1931 and 1934. They were not competitive in terms of speed, but their low fuel consumption reduced the time lost to refueling stops, allowing them to finish thirteenth and twelfth, respectively.

## THE PEOPLE'S CAR

If the Model T was the universal car of the early twentieth century, the Volkswagen was its successor. The idea of a "people's car" that would be more affordable than anything else available was aggressively pushed by Adolf Hitler, who was reputed to have a photograph of Henry Ford hanging in his office. He was something of a car enthusiast even though he did not know how to drive, and the Nazi government heavily subsidized the Mercedes and Auto Union race cars that dominated Grand Prix circuits during much of the 1930s.

Hitler saw highway building and the extension of automobile ownership as a way of reviving the German economy and solidifying popular

An early example of the "People's Car." Courtesy of the Library of Congress.

support for the Nazi government, and he enjoined the German automobile industry to build a car that could be sold for less than 1,000 Reichmarks (about $400), considerably less than anything available at the time. In order to meet this challenging target, the eminent automotive engineer Ferdinand Porsche came up with a design that departed substantially from accepted principles. The engine was in the rear, eliminating the need for a driveshaft and putting more of the car's weight over the driving wheels, an advantage when more traction was needed. The four-cylinder, horizontally opposed engine was air-cooled, eliminating the need for a water pump, coolant, and attendant plumbing. Springing was provided by torsion bars. The mechanical works were wrapped in a unit-construction, streamlined body that increased speed and economy by reducing wind resistance.

In 1934 Ferdinand Porsche contracted with the German government to design three prototypes, which were duly delivered in 1936. The car was initially named the KDF-Wagen (for *Kraft durch Freude*, or "strength through joy," the motto of the German Labor Front). In 1938 construction began on a state-of-the-art factory modeled after Ford's plant near London, with a planned capacity of 1.5 million cars a year. Building this factory and the cars it was to manufacture required more capital investment than had been provided by the German Labor Front. To secure the necessary funds, the Nazi government instituted a savings plan whereby aspiring owners made weekly payments of 5 marks. When their payments totaled 990 marks, they got their car. That, at least was the plan; of the 336,668 Germans who paid a total of 380 million marks ($67 million), not one ever received a car. Instead, their money went into the coffers of the Nazi government, and after the outbreak of World War II the factory employed slave labor to produce a variety of military products, most notably the V-1 missile.

This was in keeping with the long-term ambitions of the Nazi government, because the Volkswagen had been designed from the outset with an eye toward military applications. More than 50,000 VW-derived Kuebelwagens ("bucket cars") were produced for wartime use, where they showed themselves to be particularly well suited to desert warfare in North Africa. Allied bombing did substantial damage to the VW factory, but by August 1946 the rebuilt and refurbished facility had turned out nearly 8,000 cars. The British occupation forces suggested that the factory be turned over as part of German war reparations, but a commission of British automotive experts found its product to be hopelessly lacking in appearance, performance, and comfort, concluding that "to build the car commercially would be a completely uneconomic enterprise"(quoted in Flink 1988, 321). A similar opinion was offered by Ford's president, Ernest Breech, who reportedly told Henry Ford II that "I don't think what we are being offered here is worth a damn" (quoted in Flink 1988, 321). As we shall see, these assessments proved to be considerably wide of the mark, to put it mildly.

## AUTOBAHNS AND FREEWAYS

In Germany the Volkswagen, or People's Car, was half of the equation for mass motorization; the other half was a planned network of high-speed roads on which they could travel. Although construction of high-speed, limited-access highways predated Hitler's coming to power in 1933, it fit in perfectly with Nazi goals and ideology. Some of the motivation for building autobahns lay in the military realm, but later events showed that they were of limited value for the movement of men and material. In a cultural sense, autobahn construction was seen as an embodiment of "Nazi ideals of national character, spirit, strength, and beauty" (Lay 1992, 98), but the most important motive for autobahn construction was the alleviation of unemployment. In this it was successful, as road-building projects employed considerably more people than Hitler's aggressive rearmament program. When construction stopped in 1942, 3,200 miles (4,000 km) of autobahns were in operation.

The United States was much slower to embrace the concept of intersection-free highways. The early part of the century saw some enthusiasm for parkways, such as the Long Island Motor Parkway, which began as a racetrack for the Vanderbilt Challenge Cup, and then opened as a public toll road in 1908. Its engineering significance was that it was the first public road to use overpasses to avoid intersections, as well as the first to use banked curves.

The idea of improving the flow of traffic by replacing intersections with overpasses was revived in the 1930s as highway design began to accommodate automobiles that had much higher cruising speeds than those of a decade earlier. The first modern high-speed highway in the United States was the Pennsylvania Turnpike, which was built from 1938 to 1940, and connected Harrisburg and Pittsburgh, a distance of 168 miles (270 km). Construction costs were lowered by taking advantage of an abandoned railroad right-of-way, but they were still beyond the capabilities of the state of Pennsylvania. But as in Germany, the Roosevelt administration recognized that highway construction was an excellent way of alleviating unemployment, so 40 percent of the cost of building the turnpike were covered by a New Deal agency, the Public Works Administration, while the remainder came from a quasi-governmental agency, the Reconstruction Finance Corporation. This was not an isolated case; the New Deal era was one of greatly increased federal support of highway construction, from the previous 10 percent of national expenditures to 40 percent and more (Rae 1965, 138).

The Pennsylvania Turnpike had an element of self-financing in that it was a toll road. An alternative was to use higher gasoline taxes to finance construction costs. This was the concept underlying the "freeway," although the term also came to mean a highway free of traffic-impeding intersections and cars entering from side streets. The first Southern California freeway was the Arroyo Seco Parkway (later renamed the Pasadena Freeway), which opened in 1940 and connected Los Angeles with the neighboring city of Pasadena.

Planning for a nationwide network of high-speed roads began as early as 1941 with the formation of the National Interregional Highway Committee. This was followed in 1944 by passage of the Federal Aid Highway Act, which authorized $1.5 billion in matching funds for three years of postwar highway construction and improvement. Although representing a step in the right direction, it ended up being only a small down payment on what was to be the greatest public works project in history, the interstate highway system, one of the topics of the next chapter.

## ORGANIZED LABOR AND THE AUTOMOBILE INDUSTRY

As noted in the previous chapter, Henry Ford was widely acclaimed as a benefactor of labor when he instituted the $5-a-day wage in 1914. By the 1930s, however, his company had become notorious for its heavy-handed treatment of its workers. Next to Ford himself, the most powerful man in the company was Harry Bennett, the head of the innocuously titled Ford Service Department, the successor to the Ford Sociological Department. The key task of the Ford Service Department, which employed more than 3,000 men and has been characterized as having the largest secret police force outside Nazi Germany and the Soviet Union, was to keep Ford employees in line. Workers were subjected to a host of petty regulations; they were forbidden to talk with fellow workers, sing, or even whistle while on the job. Bathroom breaks were strictly monitored, which was facilitated by having open toilet stalls. A network of spies diffused through local communities to gather information about Ford workers.

Above all, Ford was determined to prevent his workers from organizing a labor union. In one event that gained nationwide attention, on May 26, 1937, members of the Ford Service Department attacked Walter Reuther,

Richard Frankensteen, and several dozen other union organizers as they attempted to pass out leaflets by the main gate of the River Rouge plant. The resulting "Battle of the Overpass" left several union men badly beaten. The photographs of the bloodied visages of Reuther and Frankensteen bore witness to the brutality of the Ford Service Department, but the public was still slow to abandon its perception of Henry Ford as a great benefactor of working people.

Conditions at other automobile firms were not much better than they were at Ford. Wages fell for workers fortunate to keep their jobs at a time when employment opportunities were scarce. For some workers, the answer to their difficulties was collective bargaining, the process whereby a labor union represents the workers in negotiations with an employer over wages and working conditions. Collective bargaining and the legitimization of labor unions were given a great boost in 1935, when Congress passed and President Roosevelt signed the National Labor Relations Act (commonly referred to as the Wagner Act). Among its provisions was the establishment of a National Labor Relations Board to oversee elections that would determine how workers were to be represented, and to prevent "unfair labor practices," such as firing workers who belonged to a union.

Prior to the 1930s, the American labor movement largely centered on skilled workers who belonged to unions affiliated with the American Federation of Labor (AFL). The AFL was based on "horizontal" organization; that is, its member unions included all of the workers with a particular skill irrespective of the industry or firm for which they worked. Coincident with the Depression came the rise of a new kind of labor organization that went under the banner of the Congress of Industrial Organizations (CIO). The unions belonging to the CIO were "vertical" unions; their members included all the blue-collar workers in a given industry. Thus, the United Auto Workers Union (UAW), a CIO affiliate, attempted to organize all of the workers in the automobile industry, everybody from skilled tool-and-die makers to semiskilled assembly line operatives.

The UAW was founded in 1935, initially as a constituent of the AFL. It joined the CIO the following year while retaining its affiliation with the AFL, but it was soon expelled because it refused to cut its ties to the upstart CIO. To have any relevance, the union had to organize workers at one of the Big Three firms, General Motors, Chrysler, or Ford. Although Chrysler was deemed the employer most sympathetic to workers, the leadership selected GM because it was the industry's largest firm. The union had led successful strikes at several supplier firms in late 1936, and as that year drew to a close UAW strikes closed down several GM plants. By the first week of the new year, 112,000 of GM's 150,000 production workers were on strike.

Richard T. Frankensteen, UAW organizational director, with coat pulled over his head, is pummeled at the gate of the Ford River Rouge plant in Dearborn, Michigan, May 27, 1937. This marked the first outbreak of violence in which sixteen were injured. AP Photo.

When they went on strike against GM, auto workers did not simply walk off the job. Instead they did just the opposite; they remained at their place of work. This was called a sitdown strike, and it prevented management from using strikebreakers to keep production going. The striking workers remained in the factories for six weeks, during which time they were supplied with food and other supplies by family members and local people sympathetic to their cause. One attempt to cut off the provisioning led to a violent confrontation between workers on the one side and GM security men and the local police on the other. The dangerous situation led Frank Murphy, the governor of Michigan, and Frances Perkins, the secretary of labor in the Roosevelt administration, to call for negotiations between the union and GM management. Finally, on February 11, 1937, forty-four days after the first plant was struck, General Motors agreed to allow the UAW to represent its 40,000 members (but not other GM workers) in the seventeen plants that had been shut down.

Chrysler granted recognition to the UAW two months later, but Ford was a more difficult case. A series of sitdown strikes began in March 1941, and the gigantic River Rouge plant was shut down a month later. In June the company capitulated, due in part to a threat by Henry Ford's wife, Clara, to leave him if he did not make peace with the union. The UAW now had the right to engage in collective bargaining throughout the auto industry, and it would be a formidable force in the shaping of labor relations in the years to come.

## TRAILERS AND DRIVE-IN MOVIES

Difficult decade though it was, there was more to the 1930s than economic depression and labor strife. One bright light in the era's automotive culture was the emergence of the travel trailer as a means of recreation. During the previous decade, the car-oriented campground and then the motel had encouraged long-distance automobile travel. For travelers that wanted a self-contained experience, it now became possible to hitch to one's car a trailer that offered a bedroom, kitchen, and even bathroom. Trailers were not an invention of the 1930s, but they became more feasible as roads improved and car engines became more powerful. Although the Depression severely lowered personal incomes, many people were able to acquire trailers, often building their own according to plans in the "how-to" magazines that flourished in the 1930s. For those unable or unwilling to build their own trailers, a new industry had sprung up to meet their needs; by 1937 nearly 400 firms were manufacturing on the order of 100,000 trailers a year. The most innovative of these was the Airstream Company, which employed materials and techniques that had been pioneered by the aircraft industry. Its lightweight and aerodynamic trailers are still built according to the same basic principles that had been introduced in 1935.

For those who remained at home, a new automobile-centered industry began to serve people's recreational needs: the drive-in movie. The idea of projecting a movie on an outdoor screen to be viewed in one's car originated with Richard Hollingshed, who received a patent on the drive-in movie in 1933, the same year that he opened the first example in Camden, New Jersey. In this new kind of theatre, cars were parked on a gently sloped plot of land so the occupants had a good view of the screen while they listened to a soundtrack that emanated from speakers housed in the same building as the projector. Residents of neighboring homes and businesses found this annoying, and the drive-in's patrons were distracted by the lack of synchronization of sight and sound caused by the different speeds of the two. These problems were overcome in the 1930s through the supply of individual in-car speakers. Hollingshed's patent was overturned by the U.S. Supreme Court in 1938, but this did not unleash a wave of drive-in construction. As we will see, the industry did not really take off until a decade later, when demographic and social changes powered its rapid growth.

## "A GREAT ARSENAL OF DEMOCRACY"

World War I had been a war of attrition in which a superior ability to supply industrialized armies tipped the balance in the favor of the Allied powers. World War II was no different. Although Hitler pinned his hopes for the domination of Europe on blitzkrieg ("lightning war"), the war in Europe dragged on for nearly six years. When it finally concluded in 1945, the superior productive power of the Allies, and that of the United States in particular, was the decisive force in thwarting German and Japanese aggression.

No sector contributed more to the war effort than the automobile industry. Not only was the United States far and away the most advanced car-producing nation, but its ally Great Britain made more cars in 1938, the last year of peace, than Italy, Germany, and Japan put together. In the years immediately preceding the outbreak of the war, the British government had sponsored a "shadow factory" plan whereby it constructed aircraft engine factories operated and managed by the automobile industry. The first engines were produced in shadow factories at the end of 1938, and beginning in 1940 a second generation of these factories went into operation, building aircraft engines and even complete aircraft. Across the Atlantic, Britain's Dominion partner, Canada, served as a major contributor of trucks and other equipment to the war effort.

Although isolationist sentiment was strong in the United States in the years immediately prior to the outbreak of the war, the prospect of American involvement could not be ignored. In recognition of the need to begin planning for wartime production, President Roosevelt formed the National Defense Advisory Council to organize military production. Significantly, the council was headed by William Knudson (1879–1948), who had resigned from the presidency of General Motors to take the new post. In contrast, Henry Ford, transfixed by the delusion that Roosevelt was in league with General Motors and the du Ponts to get control of his firm, very reluctantly agreed to gear up for aircraft engine production.

The Japanese attack on Pearl Harbor put a swift end to isolationist sentiments, and was followed by rapidly putting the economy on a wartime footing. The automobile industry was one of the first to make the transition, as the manufacture of vehicles for the civilian market came to an end on February 22, 1942. Car travel was severely curtailed by the rationing of gasoline and tires, and even when these were available, motorists were restricted to a nationwide top speed of 35 mph, which was primarily intended to reduce wear on tires. Everything was now geared to the war effort, and the automobile industry responded in a most impressive fashion. Through the course of the war, the industry turned out seventy-five essential products with a total value of $29 billion, constituting one-fifth of the nation's output of military goods. As might be expected, engine production was prominent; the American automobile industry manufactured 4,131,000 engines, of which 450,000 were complex aircraft power plants, along with 27,000 airplanes (Flink 1988, 276). The industry also accounted for 47 percent of the machine guns, 57 percent of the tanks, 85 percent of army helmets, and 87 percent of the bombs (Rae 1965, 158).

Of all of the industry's aircraft manufacturing operations, the most ambitious was Ford's effort to produce the most complex airplane of its time, the B-24 Liberator bomber. The firm had already played an important role in aviation history through its production of the Ford Trimotor commercial airliner, and Henry Ford was convinced that established airplane manufacturers needed an infusion of his superior knowledge of mass production, which was underscored by his wild boast in 1940 that he could turn out 1,000 planes a day if he could avoid government "meddling." Ford soon discovered that building airplanes in large quantities was far more difficult than turning out millions of cars and small trucks. Not only were military aircraft far more complex, but it was also often necessary to make substantial modifications to them as a result of lessons learned in combat.

Building the bombers entailed the construction of a new plant at Willow Run near Ypsilanti, Michigan (this led skeptics to refer to the project as "Will it run?"). Under the leadership of Charles Sorensen, the Willow Run plant introduced a number of useful productive innovations, such as building the airplane's fuselage in two longitudinal halves to facilitate the installation of wiring. Eventually the plant was turning out one airplane every hour, and it made nearly 7,000 planes in all, but by the time it did so the B-24 was being supplanted by the larger and faster Boeing B-29. It is thus not evident that Ford's massive involvement in aircraft production justified the tremendous expenditures and efforts that went into it.

Important as the production of aircraft was, the industry could not slight its core business, motor vehicles. Although only a few hundred automobiles were produced for use as staff cars, the industry made vast numbers of trucks, 2.5 million of them, the majority equipped with four-wheel drive. Four-wheel drive was also a key feature of one of the war's most significant vehicles, the Jeep (derived from GPV, general purpose vehicle). Created by Captain Robert G. Howie in response to an army competition for a small armed vehicle, and first exhibited at Fort Benning, Georgia, in 1940, the Jeep was originally produced by the American Bantam Car Co., a firm that had been manufacturing small numbers of tiny cars based on the English Austin 7. The new vehicle's merits were evident, but Bantam was too small to make them in the required numbers, so the bulk of the 660,000 Jeeps built during the war were manufactured by Ford and Willys-Overland Motors, with the latter firm continuing to make them after the war as the progenitor of today's SUV.

The impressive production record of the American automobile industry would have been impossible without the great efforts of its workers, many of whom had little or no industrial experience prior to the war. African Americans and other minorities, retirees, students, and women all made major contributions to the war effort. "Rosie the Riveter," the personification of the female industrial worker, became a national icon that continues to inspire later generations.

In 1945 an icon of a different sort left the automotive scene. Henry Ford, always headstrong and eccentric, had lapsed into senility after suffering a series of strokes between 1938 and 1941. His son Edsel, who had been a steadying influence at the firm, died at the age of 49 in 1943, making Harry Bennett the main force at Ford. At this point, concerns about the increasingly chaotic situation at Ford and its effects on wartime production caused some government officials to contemplate a government takeover of the company. Instead, Edsel's oldest son, Henry Ford II, received an early discharge from the U.S. Navy to become executive vice president in 1944. Bennett still hoped to take over the company after the founder's death, but Eleanor Ford, Edsel's widow, threatened to sell the 41.9 percent of Ford stock that she owned if her son was denied the presidency of Ford. This tactic was effective, and in 1945 Bennett was out, and Henry Ford II took charge of the firm founded by his grandfather.

The death of Henry Ford two years later and of Billy Durant in the same year drew the curtain on an era. Two of the industry's giants had departed, but the principles of manufacturing and marketing that they had pioneered remained in full force. The following years would see a great expansion in automobile production in the United States and in many other countries, accompanied by a variety of economic and social changes as the world continued to adapt to an automobile-driven culture.

5

# In High Gear, 1945–1965

◆

Economic depression and global warfare had severely strained the automobile industry and the motoring public. But bad memories could be put behind as automobile ownership expanded rapidly amidst the relative peace and prosperity of the post–World War II era. A plethora of new models lured buyers in Europe and the United States. The cars produced in the two continents continued to diverge in their basic engineering and style, yet by the late 1950s growing numbers of Americans were discovering the virtues of cars imported from abroad, and by the end of the period Japan had unexpectedly emerged as a major automobile producer and was on the threshold of becoming a large-scale exporter of automobiles and small trucks. Complementing the relentlessly increasing number of foreign and domestic cars was an expanding highway system that required massive involvement of the federal government in road construction and maintenance. These highways played an essential role in making possible some of the key social transformations of the postwar era, most prominently the seemingly unstoppable spread of suburbia, but also including the proliferation of franchised motels and fast-food restaurants throughout the land.

## THE RETURN OF MASS MOTORIZATION

When World War II ended with Japan's surrender on August 14, 1945, the prospects for the postwar economy were not bright. The years leading up to the war had been marked by economic stagnation and widespread unemployment, and many in the United States thought that without the stimulus of military spending the economy would soon revert to this unhappy situation. Making matters worse, millions of men and women were being released from military service and were suddenly injected into the labor market just as wartime production was grinding to a halt. Prices, now freed from government controls, began to rise at a rapid clip. Leaders of labor unions were demanding higher wages, and backed up their demands with threats of crippling strikes.

Although the United States had been spared the material destruction suffered by the war's combatants, anxiety over a return to Depression-era conditions replaced fears of military defeat. In much of Europe and Asia, the situation was far worse; for many of its people the effort just to get through the day left little time or energy to think about the future. In Britain, the enormous cost of winning the war was all too evident. The treasury was empty, many goods continued to be rationed, and exports took precedence over meeting domestic needs. For the losers, the situation was desperate. German and Japanese cities had been devastated by Allied bombing, the industrial infrastructure was crippled, and daily hunger was an all-too-common experience for many people.

Under these circumstances, few would have foreseen a postwar boom in automobile ownership. Yet in the decade immediately following World War II, an economic miracle unfolded, and automobile production reached unprecedented levels. By 1965, a scant two decades after the end of World War II, an automobile-based culture had taken hold in the industrialized world. In sharp contrast to the grim predictions of the immediate postwar period, Americans, Europeans, and Japanese were enjoying an unprecedented rise in their personal incomes, and a significant proportion of this newfound prosperity was closely tied to a rise in automobile ownership. A vicious circle had been transformed into a virtuous circle. During the 1930s the collapse of the economy went hand in hand with a precipitous drop in automobile production. Now the situation was reversed; a thriving economy and a booming automobile industry combined to produce an economic boom of unprecedented proportions.

If the collapse of the 1930s was in part due to the automobile industry having more capacity than could be absorbed by the car-buying public, the opposite situation prevailed during the second half of the 1940s. Due to the suspension of production during World War II, the majority of the cars on the road dated back to the 1930s or earlier, and most of them were worn out after years of hard use. Consumers were desperate for new cars, and the marketing of prewar designs encountered no sales resistance. The initial problem, therefore, was converting from the production of war materiel to the manufacture of automobiles.

As the only major participant in World War II not to suffer physical damage, the United States quickly returned to its accustomed place as the world's largest producer of automobiles, accounting for 82 percent of world output in 1950. America did not hold so dominant a position for long. In 1950 the United States could claim twenty-six cars for every hundred people, while France and West Germany had a mere four cars per hundred people, and Britain was only slightly better with five cars for every hundred. But as Table 5.1 shows, the ownership of cars in Western Europe expanded rapidly, and by 1970 the ratio of cars to people was almost equivalent to the United States in 1950.

Automobile manufacturing was one of the industrialized world's great growth industries during the first two decades of the post–World War II era. In 1951 total output was fewer than 7 million cars. By 1960 the number was up to nearly 13 million, and by the time the decade ended, 22.7 million cars were rolling out of the world's automobile factories, a more than threefold increase (Sedgwick 1983, 37).

In the first postwar decade, Britain was Europe's largest producer of automobiles, and it stood second only to the United States in total production. In 1949 it was the world's largest exporter of automobiles, due in part to a government policy that set export quotas for individual manufacturers and enforced them through the allocation of steel. This was a position that could not be maintained. Although they were well matched to their home market, British cars were not suited to places with poor roads, long travel distances, and hot climates. Much more successful over the long run were the makers of German cars. Led by Volkswagen, Germany surpassed Britain

## Table 5.1
## Automobiles per 100 Population, 1960 and 1970

|  | 1960 | 1970 |
|---|---|---|
| United States | 34 | 43 |
| France | 11 | 24 |
| West Germany | 9 | 23 |
| Great Britain | 11 | 21 |
| Italy | 3 | 19 |

Adapted from Bardou et al. 1982, 197.

as the world's top exporter in 1955, and it passed Britain in total production the following year.

France's largest manufacturer, Renault, followed the Volkswagen pattern with the introduction in 1946 of its rear-engine 4CV, although it had four (tiny) doors and a water-cooled engine. More idiosyncratic was the Citroën 2CV that first appeared in 1948. Powered by a 375-cc, two-cylinder, air-cooled engine, it was as utilitarian as a car could be. According to a frequently told story, the original design specification required that the car had to be able to cross a freshly plowed field without breaking any of the eggs in a basket perched on the seat.

England, France, and Germany had long histories as suppliers to motorists everywhere. In contrast, Sweden's car industry had largely served a domestic clientele. Beginning in the 1950s, however, it emerged as a serious force in the world market. The leader was Volvo; its PV544 sedan was thoroughly conventional down to its styling, which resembled a shrunken Detroit product from the 1940s, but it was well-made and reliable, and it enjoyed considerable sales success in the United States. A more distinctive Swedish design was the SAAB. Introduced in 1949, its aerodynamic form reflected its origins as the product of an aircraft company (SAAB stands for Svenska Aeroplan AB, or Swedish Aircraft Company). Its running gear owed a great deal to the prewar German DKW, with a two-stroke engine (initially with two cylinders and then three) that drove the front wheels.

The Netherlands had lacked a native automobile producer for many years prior to and immediately after World War II. The Dutch automobile industry reemerged in 1958 with the production of the DAF. A small car powered by a two-cylinder engine, its unique feature was an automatic transmission built around rubber belts and pulleys.

While Western European automobile manufacturers were rolling out a host of innovative designs in ever-increasing quantities, little of interest was being produced by the Soviet Union and its satellite countries, which relied heavily on obsolescent technologies that had been developed elsewhere, as was the case with Russia's Opel-derived Moskvitch, and the East German Trabant, which like the SAAB was based on the two-stroke engine and front-wheel drive technology of the prewar DKW.

## MORE MODELS, FEWER MAKERS

The great expansion of automobile production during the 1950s was paralleled by a proliferation of car models. To take one example, in 1951

Chevrolet produced one basic car body and used the same six-cylinder engine for everything it sold. Only trim and a few details separated the different model lines. By 1959 there were no fewer than eight different models of cars bearing the Chevrolet badge, and these were powered by everything from a 152-cubic-inch (2.5-liter) four-cylinder to a 427-cubic-inch (7-liter) V-8 engine.

Paradoxically, the increase in the number of car models was paralleled by a decline in the ranks of automobile manufacturers. The sellers' market of the immediate postwar years had given some breathing room to smaller manufacturers; in the long run, however, they had little hope of successfully competing with the industry giants. Yet while they lasted they enlivened the industry with dramatically styled products, such as the futuristic Studebaker and the "step down" Hudson that seated its occupants lower to the ground than any other car in production at the time.

The immediate postwar years also saw a few new firms attempting to take their place alongside the long-established manufacturers. The most romantic saga is that of the Tucker, which even became the subject of a popular movie long after the firm's demise. The creation of Preston Tucker, an entrepreneur with many years of experience in the automobile industry, the Tucker car embodied a number of novel features like a rear-mounted, air-cooled, flat-six engine and an auxiliary headlight that pointed in the direction that the car was being turned. By 1950, however, undercapitalization and consequent difficulties in getting the car to market put an end to the firm after only fifty-one examples of the Tucker Torpedo had been built.

Much more in the automotive mainstream were the cars made by Kaiser-Fraser. Henry Kaiser, the firm's cofounder, had won renown during World War II for his company's use of mass-production techniques for the rapid construction of Liberty Ships. Joseph Fraser, the other part of the firm's name, was a respected automobile industry executive. If any new firm could gain a permanent niche in the U.S. automobile industry, surely this was the one. Although Kaiser-Fraser built a prototype front-wheel-drive car, its production models used conventional rear-wheel drive. In 1947 and 1948 Kaiser-Fraser was the eighth largest selling make in the United States, and the biggest of the independents with 144,490 and 181,316 cars produced in these two years. This, however, was the high-water mark. Sales plummeted to 57,995 in 1949 as the major manufacturers came out with their first fleet of all-new models. A restyled body for the 1951 model year failed to halt the slide in sales, and 1955 marked the last year of production.

The mid-1950s were also unkind to long-established automobile manufacturers outside the Big Three (General Motors, Ford, and Chrysler), who by this time held 96.4 percent of the U.S. market (Bardou et al. 1982, 176). Hudson and Nash merged in 1954 in an attempt to stay alive as sales hit the skids, but this only postponed the day of reckoning until 1957, when the last cars bearing Nash and Hudson nameplates left the factory. However, the merged firm, now known as American Motors, enjoyed a number of years of success with its Rambler line of economy cars. Less successful was the merger of Studebaker and Packard, which was also consummated in 1954. The last true Packard was built in 1956, and slightly restyled Studebakers were sold as Packards until 1958. Studebaker was revived by the success of its compact Lark, but this was only a temporary reprieve, and the last Studebaker was built in 1966. The American automobile industry had been reduced to the Big Three and one "independent," American Motors.

## EUROPEAN AUTOMOBILE DESIGN

Throughout the 1950s and 1960s, European manufacturers responded to an expanding market with a host of new models that embodied a great variety of technological approaches to individualized transportation. Much of that innovation was a response to governmental policies and to the distinctive driving conditions found in Europe. Britain scrapped its "horsepower" tax in late 1946, finally removing the incentive to produce cars with small-displacement, long-stroke engines, but high taxes on gasoline mandated the continued production of small cars with engines to match. Britain was not a unique case; fuel taxes in Europe continued to be much higher than they were in the United States, impelling manufacturers to offer space-efficient cars powered by engines that were tiny by American standards. In France, cars with large-displacement engines also were subjected to very high taxes. Faced with a sharply contracting market, illustrious makes like Delahaye, Talbot, and Delage went out of business.

Efforts to maximize interior space relative to overall size led many manufacturers to "put the works at one end," by combining the engine, gearbox, and differential in one compact package. Germany's Volkswagen set the pattern, which was echoed by the Renault 4CV and its successor the Dauphine, which was introduced as a 1956 model. The rear-engine brigade also included the German NSU (1957), the British Hillman Imp (1963), and the Italian Fiat 500 (1957) and 600 (1955). Front-wheel drive also had its adherents in the German DKW, Swedish SAAB, and Citroën DS-19, but as we shall see it was not until the late 1950s that its full potential was realized.

## ENGINEERING AND DESIGN:
## THE UNITED STATES

While the Europeans were building small cars with fuel efficiency as a key design criterion, U.S. automakers were going in the opposite direction. Much of the technological innovation of the late 1940s and early 1950s centered on the development of increasingly powerful engines with prodigious appetites for gasoline. The future trajectory of American automobile engines had been set by Ford in the 1930s. Although its cars were technologically stagnant in a number of ways (Ford stuck with mechanical brakes until the 1939 model year and solid-axle front suspensions for ten years after that), the performance offered by its V-8 engine helped it to remain competitive with its domestic rivals. The appeal of a smooth, powerful V-8 engine was not lost on General Motors, and for the 1949 model year its Cadillac division offered the most powerful engine in an American production car since the demise of the Duesenberg. Unlike the flathead Ford, the Cadillac V-8 employed overhead valves, which increased volumetric efficiency by improving gas flow during the intake and exhaust phases. The pushrods, rocker arms, and other valve train parts made the engine more expensive to manufacture, but this was not a serious problem for the most costly car in the General Motors line. Displacing 331 cubic inches, the Cadillac V-8 developed 160 hp. Although this represented only 10 more horsepower than the flathead eight that it replaced, the engine produced more torque, improved fuel economy by 14 percent, and was 200 pounds lighter.

In the same year that the Cadillac V-8 appeared, Oldsmobile, which had a reputation as GM's "experimental" division, offered its own V-8. Displacing 303 cubic inches (5.0 liters) it produced 135 hp. Both Cadillac and Oldsmobile engines had the potential to produce considerably more power, and their compression ratios were increased as gasoline with higher octane ratings began to be marketed, resulting in a significant power increase. Through the 1950s and beyond, the two GM divisions also repeatedly resorted to the tried-and-true method of increasing power by increasing bore and stroke. By 1956 Cadillac and Oldsmobile engines displaced 365 cubic inches (6.0 liters) and 324 cubic inches (5.3 liters), and produced 305 hp and 240 hp, respectively. By 1965 the numbers were 429 cubic inches (7.0 liters) and 340 hp for Cadillac, and 455 cubic inches (7.5 liters) and 390 hp for Oldsmobile. (It should be kept in mind that the horsepower figures supplied by the manufacturers reflected optimal test procedures and were much higher than the power actually delivered to the wheels.)

For the 1951 model year, Chrysler went General Motors one better by bringing out an overhead-valve V-8 with hemispherical combustion chambers, the legendary "Hemi." A hemisphere offers the theoretically optimal shape for getting the most air-fuel mixture into the combustion chamber, and the Chrysler engine soon gained a reputation for having the greatest power potential. In its initial form, the Hemi had a displacement of 331 cubic inches (5.4 liters) and produced 135 hp, and as with its GM rivals, it grew in size and horsepower through the decade; by 1957 it was possible to get a stock Chrysler with as much as 390 hp under its hood. In the hands of drag racers, the engine could be induced to give over 1,000 hp for brief runs down a quarter-mile strip.

The Hemi was a complex engine that powered an expensive line of cars (although Dodge got a smaller version of the Hemi in 1953). At the other end of the price scale, General Motors' Chevrolet division introduced its own V-8 for the 1955 model year. It, too, was destined to achieve legendary status even though it was far cheaper to build than the Hemi. Designed with manufacturing economy in mind, the Chevrolet displaced 265 cubic inches (4.3 liters) and produced 162 hp, and 195 hp when equipped with a four-barrel carburetor, dual exhausts, and a high-performance camshaft. As with the other V-8 engines, its displacement was increased over the years and the power it put out was increased substantially through the use of fuel injection and other performance enhancements. The so-called small block V-8 has been the basis of many racing engines, and its long life as a car and truck power plant is underscored by the fact that more than 20 million of them have been built since the engine's introduction fifty years ago.

Powerful V-8 engines formed a natural complement to the other major American-led technological advance of the late 1940s and early 1950s, the automatic transmission. With their elaborate innards of fluid couplings, gears, and brake bands, automatic transmissions absorbed a significant amount of power. But what would have been problematic when such a transmission was mated to an 80-hp engine was of much less significance when the power plant was at least twice as powerful. As we have seen in the previous chapter, Oldsmobile was equipped with the first successful fully automatic transmission for the 1939 model year, and after the forced hiatus of World War II, General Motors wasted little time in equipping its other makes with automatics. The wisdom of this move was soon evident. By 1950 all Cadillacs were equipped with automatic transmissions, along with 94 percent of Oldsmobiles, 77 percent of Buicks, and 65 percent of Pontiacs. Chevrolet came out with a less sophisticated two-speed transmission, Powerglide, for the 1950 model year. Twenty percent of buyers opted for it, and that percentage doubled with the next model year (Sedgwick 1992, 204). Other manufacturers had little choice but to follow suit, even if it meant buying their transmissions from General Motors.

By the time V-8 engines and automatic transmissions had become ubiquitous in American cars, the United States had been a firmly established automobile culture for four decades. Automobile ownership was an integral part of the lives of tens of millions of people, and the majority of them had little interest in the mechanical details of the cars they drove, while driving conditions in many parts of the country made few demands on motorists' skills. Under these circumstances, automatic transmissions fit in perfectly with a rather uninvolved approach to driving; all that was necessary was to press down on the accelerator and brake pedal at the appropriate times, and to use the steering wheel to point the car in the general direction that one wanted to go.

While engines and transmissions absorbed the energies of American automotive engineers, the rest of the chassis remained technologically stagnant. Decades of improvements had made roads and highways straighter and less demanding, so all that most drivers required of their cars' suspensions was that they provide a smooth ride and insulate them and their passengers from the world outside. This was achieved with no fundamental changes to suspension design, except for Packard and Chrysler's use of torsion bars as a springing medium. Cadillac (along with Borgward and Mercedes in Germany) offered air suspension on some of their cars, but problems with air leakages led to its rapid withdrawal from the market.

In similar fashion, brake technology was neglected, with potentially lethal consequences for the drivers of American cars. Hydraulically actuated drum brakes were adequate when automobiles were relatively light, and cruising speeds were in the neighborhood of 60 mph. But by the end of the 1950s, many American cars were tipping the scales at well over 4,000 pounds and were capable of comfortably cruising at over 80 mph. Making matters worse was the use of smaller brake drums that resulted from stylists' efforts to make cars lower through the use of smaller wheels. Under these circumstances, drum brakes were not up to the task of effecting a controllable stop within a reasonable distance. Overburdened drum brakes also exhibited a pronounced vulnerability to "fade"; after a few hard applications, the brakes became so hot that thermal expansion of the brake drums increased the clearance between them and the brake shoes, with a consequent loss of effectiveness. Under these circumstances, driving down a mountain road could be a frightening experience. In the 1950s, manufacturers introduced power assistance, but it did not alleviate the inherent weakness of undersized drum brakes with inadequate cooling. Power brakes lowered the pedal effort, but this only disguised a dangerous lack of stopping power.

Much better braking performance was offered by disc brakes, which were originally developed in the aviation industry, where stopping aircraft with high landing speeds had been a serious problem. Precursors to disc brakes can be found in the early twentieth century, but it was not until the 1950s that they were effectively used for automobiles. In the United States, disc brakes were used as early as 1949 on a few Chryslers and on the tiny Crosley, but these were strictly low-production efforts. As with many of the automotive innovations of the postwar era, European manufacturers set the pace. The superior performance of disc brakes was brought to public attention when Jaguar used them on its Le Mans–winning race cars from 1952 to 1955. In the latter year, Citroën fitted disc brakes to its radically styled and engineered DS19, and Triumph began to use disc brakes the following year. In both cases discs were fitted only to the front wheels, which provide most of the braking power due to forward weight transfer under deceleration. Other European makes followed in short order, but with the exception of the 1962 Studebaker Avanti and the 1963 Corvette, disc brakes did not become available on American cars until the mid-1960s, and even then they were usually made optional at extra cost.

European firms were responsible for another important contribution to improved automotive safety and performance, the radial tire. Unlike bias-ply tires that were reinforced by diagonal fabric plies, radials used steel cords running at right angles to the tire's rim. This method of construction gave the tires strong yet flexible sidewalls, resulting in better road adhesion, a greatly reduced tendency to hop around during fast cornering, and a much longer lifespan than conventional bias-ply tires. Michelin in France and Pirelli in Italy did much of the early development work on radial tires, which were fitted as standard equipment on many European cars after their introduction in 1947. American manufacturers were much slower to adopt them, in part because they resulted in a slightly harsher ride, and during the 1950s and 1960s, a smooth ride was of supreme importance for American automobile chassis engineers.

While soft suspensions were providing at least superficial riding comfort for the occupants of American cars, air conditioning made for a much pleasanter experience when traveling in hot weather. Automotive air conditioning first appeared on luxury cars, the earliest examples being fitted to a few 1940 Packards, where they added about 25 percent to the purchase price. Cost and weak performance initially limited the market for automotive air conditioning; only 10,500 air-conditioned cars were sold from 1939 to 1953, and as late as 1965 only 10 percent of the automobiles sold in the United States had air conditioning. The slow emergence of automotive air conditioning reflected the inherent difficulty of adapting air conditioning to automotive use. The space to be cooled in a car's interior is much smaller than that of a home or office, but so is the space for installing the necessary equipment. Moreover, a car's air conditioner has to be able to cool the interior much more rapidly than one used for a building, and car interiors get much hotter after being parked in the sun. These challenges were eventually met as advances in the design of compressors and other components improved performance and brought down costs. By the early 1990s well over 90 percent of the cars sold in the United States were equipped with air conditioning.

While the comfort and straight-line performance of American automobiles made impressive strides forward, far less attention was being paid to making them safer. Racing experience indicated that deaths and serious injuries could be reduced when drivers were firmly strapped in their cars, but manufacturers were slow to fit their products with seat belts. Nash equipped some of its cars with seat belts in 1952, while a more comprehensive approach to safety was taken by Ford for the 1956 model year, when it offered seat belts (for an additional charge of $9) along with a number of other safety features: breakaway rearview mirrors, crashproof door locks, and steering wheels with recessed hubs. For an additional $16, buyers also could have padded dashboards and sun visors. It is often said that these safety improvements were met with cold indifference by the buying public, and in fact only 20 percent of the 1956 Fords were so equipped. But this figure compared reasonably well with the percentage of cars equipped with other novel features, such as automatic transmissions, when they first appeared. In any event, the issue of automotive safety had only barely emerged, but it would become a dominant theme by the middle of the following decade.

## FINS, CHROME, AND TUTONE PAINT JOBS

With safety, braking power, and suspension design receiving scant attention, there was little to set one make of American car apart from another. Automobile manufacturers reckoned that the only thing that distinguished their respective products was their appearance. Much of General Motors' past sales success was attributed to its leadership in automotive style, a trend that started in the 1920s with the establishment of Harley Earl's Art and Colour Section. Led by GM, the overriding theme of American car design during the late 1940s and for all of the 1950s can be summarized as "longer-lower-wider." The results of this philosophy can be seen in a comparison of the measurements of a 1949 Chevrolet, the division's first postwar design, and the 1965 version (Table 5.2).

Complementing the visual sleekness of the new generation of automobiles was a styling idiom that borrowed heavily from the great technological icon

**Table 5.2**
**Chevrolet Dimensions, 1949 and 1965**

|  | 1949 | 1965 |
|---|---|---|
| Wheelbase | 115.0 in. | 119.0 in. |
| Length | 197.0 in. | 213.0 in. |
| Height | 65.9 in. | 55.5 in. |
| Width | 73.9 in. | 79.5 in. |
| Weight | 3,125 lb. | 4,140 lb. |

of the Cold War era, military jet aircraft. Throughout much of the 1950s, automobile design was characterized by wrap-around windshields and roofs shaped to look like cockpit canopies, fake air scoops, and exhaust pipes made to look like the tail end of a jet. The 1950s were also the great age of chrome, where once again General Motors led the way. One egregious example was its 1958 Oldsmobile, which was weighed down by 44 pounds of chrome trim. More than at any other time before or since, American automobile design had become almost completely detached from the functional requirements of transporting people safely and efficiently.

In sharp contrast to GM cars, Chrysler products of the early 1950s emanated an air of practical sobriety. Boxy and with a high roofline, they underscored Chrysler's image as a no-nonsense, engineering-driven manufacturer of automobiles. Their Hemi V-8 won accolades for technical sophistication, but the cars that they powered were poor sellers; apparently the buying public wanted something other than roomy practicality. Determined to redress declining sales, Chrysler adopted a flamboyant design idiom called "the Forward Look." For the 1957 model year, Chrysler brought out a stunning line of cars characterized by long, low profiles and soaring tailfins. Fins were not unique to Chrysler; their appearance can be traced to the two small humps on the rear of the 1949 Cadillac that had been inspired by the twin booms of the Lockheed P-38 fighter. But Chrysler's fins were unmatched for size and visual appeal, and motorists voted their approval with their checkbooks.

The following year, however, was a severe letdown, as a sharp recession and a warmed-over repetition of the previous year's styling seriously cut into Chrysler sales. Even worse was the experience of the Ford Motor Company, where an effort to launch a midprice car failed disastrously. Dubbed the Edsel, the body design of the new Ford product missed the mark badly. The front end with its large vertical grille was singled out for criticism; some likened it to "an Oldsmobile sucking a lemon," while others offered anatomical analogies best not repeated here. The Edsel encountered so much sales resistance that Ford discontinued it early in the 1960 model

The Edsel embodied some of the worst excesses of late 1950s styling, most notably its controversial grille. Courtesy of the Library of Congress.

year after racking up losses estimated at $400 million, and the word "Edsel" passed into the English vocabulary as a synonym for costly failure.

The Edsel fiasco can in part be attributed to its clumsy styling, but it was hardly unique in this regard. Automobile sales in the United States fell to their lowest level in ten years as many consumers rejected the bloated and tastelessly styled 1958 models. Instead, the car-buying public was turning to smaller, more restrained designs such as the American Motors Rambler and the Studebaker Lark. Even more disturbingly, foreign cars, Volkswagens in particular, were enjoying a sales boom. In response, the Big Three brought forth a range of "compacts," the Ford Falcon, the Plymouth Valiant, and the Chevrolet Corvair.

All of the domestic import-fighters were good sellers, and for a while they succeeded in stemming the tide of European cars coming to the United States. All three gained some measure of notoriety. The Valiant was renowned for its indestructible Slant-Six engine (so named because it was canted over to allow a lower hood line). The rear-engine Corvair became the focal point for an indictment of the American automobile that has had repercussions ever since. The Falcon was a straightforward car that might have slipped into obscurity had it not provided the basic structure for one of the great marketing successes in the history of the American automobile industry, the Mustang.

Introduced as a 1965 model in April 1964, the Mustang set the pattern for a new automotive niche, the relatively small and inexpensive sporty car, a style that pays homage to its progenitor with the term "pony car." With its long hood and short deck, the Mustang gave the appearance of having a powerful engine under its hood, even though the base model, which listed for only $2,368, came equipped with a puny 101-hp six-cylinder engine. Most buyers wanted more, and the average Mustang was equipped with $1,000 worth of optional extras like a 270 hp V-8, four-speed gearbox, tachometer, and even disc brakes.

The Mustang was a runaway success, with 418,000 sold in its first year of production. Its sporty appearance and low price had a lot to do with its commercial success, but no less important was the fact that it was perfectly attuned to its era. It was during the mid-1960s that the first wave of the postwar baby boom generation was reaching maturity. The boomers made up a huge demographic cohort, and they had a lot of money to spend in comparison with the generation of the Depression and World War II. Ford's competitors were caught off-guard by the Mustang's appearance, but General Motors had similar cars under development in the form of the Chevrolet Camaro and the Pontiac Firebird that appeared in 1967. Chrysler had to resort to grafting a huge fastback roof onto the Valiant body; its feeble sales bore witness to the inadequacy of its efforts.

## THE MINI

While American cars appeared to be stuck on a technological plateau, European designers were rewriting the book on automobile engineering. As we have seen, the quest for economy and efficiency motivated designers to concentrate the powertrain at one end of the car, usually the rear. This, however, had its disadvantages. A rear-engine car provided meager luggage accommodations, and the rearward weight bias resulted in a sensitivity to crosswinds and tricky handling. In contrast, front-wheel drive offered a more space-efficient way to accommodate a car's occupants and their luggage while retaining the traction advantage of concentrating weight over the driving wheels. At the same time, however, putting power to the wheels that also steered the car presented a number of engineering challenges. Some previously mentioned cars like the Cord, various Citroëns, DKW, and SAAB enjoyed a measure of technological and commercial success, but they remained outside the mainstream of automotive design.

Front-wheel drive received a tremendous boost in 1959 with the appearance of a car that foreshadowed the basic layout of many small cars to come, and many medium-sized ones as well. The car was a product of the British Motor Corporation, which had resulted from the 1953 merger of the two largest British firms, Austin and Morris. Originally known as the Morris Mini Minor and the Austin 7, before long it was simply known as the Mini. Designed by Alec Issigonis as a response to soaring gasoline prices in the wake of Egypt's closure of the Suez Canal in 1956, the Mini maximized the interior space of a very small car. This was done by mounting the engine, gearbox, and differential at the front of the car. To further save space, the engine was installed transversely, while the gearbox lived in the oil sump directly below the engine. The tiny 10-inch (25.4-cm) diameter wheels were placed at each corner of the car, further saving space. As a result of all of these measures, 80 percent of a car measuring only 10 feet (3.05 m) in length was devoted to accommodations for the driver and passengers. The suspension also reflected an innovative approach, using rubber cones in torsion instead of steel leaves or coils as a springing medium.

By 1965 a million examples had been built, and the word "mini" had made its way into the English lexicon in a variety of guises, "miniskirt" being

Britain's Mini maximized interior space while minimizing external size. Courtesy of the MINI Division of BMW of North America, LLC.

perhaps the most memorable. Unfortunately for BMC, the technical and cultural significance of the Mini was not paralleled by financial success. BMC's top management believed that the car had to be sold at a price equivalent to its cheapest competition, even though it was far more technologically sophisticated and expensive to make. As a result, the Mini never returned a substantial profit to the company that produced it.

Although the Mini may have contributed to the eventual financial collapse of British Leyland, the successor of the British Motor Corporation, it still has to be reckoned one of the most influential cars of all time. BMC and BL used the Mini's basic layout for the larger Morris and Austin 1100 (and again lost money due to pricing them too low), as well as a number of other cars that failed to replicate the Mini's success.

While the Mini's manufacturers floundered, their car was inspiring a host of small transverse-engined, front-wheel-drive cars like the Fiat 128 and the Peugeot 204. Other manufacturers switched to front-wheel drive but retained the more conventional fore-and-aft engine location for their cars, such as the Renault 16, the Lancia Fulvia, and the Triumph 1300. In the United States, the 1966 Oldsmobile Toronado boasted the largest engine ever used in conjunction with front-wheel drive, 427 cubic inches (7 liters) and a stated 385 hp. A car of the Toronado's dimensions hardly needed the packaging efficiency offered by front-wheel drive, but it demonstrated the workability of the concept even when taken to an almost absurd extreme.

Despite the competition offered by the more advanced cars that took their inspiration from it, the Mini continued to find new customers, allowing it to remain in production until October 2000, by which point a total of 5.5 million had been built. Although it did not reach the production figures of the Ford Model T and the Volkswagen Beetle, it stands alongside them as one of the most important cars of all time, both for what it was and for the path that it blazed.

## ROTARY ENGINES

While American manufacturers were taking automotive design on new flights of stylistic fancy and European manufacturers were developing innovative ways of improving space utilization, one thing remained constant: the kind of engine found under the hood. The vast majority of the cars of this era used conventional spark-ignited internal combustion engines, the exception being the diesel-engined cars offered by Peugeot, Fiat, Borgward, Standard, and Mercedes, with only Mercedes producing them in significant numbers. Some automotive visionaries believed that this was a temporary phase, and that the conventional internal combustion engine with its inelegant use of reciprocating pistons to produce rotary motion would be supplanted by a pure rotary motion engine in the not-too-distant future.

The 1950s was a time when turbojet engines were replacing piston engines in many types of aircraft, so why couldn't the same thing be done with automobiles? A few firms succeeded in building prototype turbine-powered cars, and in 1964 Chrysler took the concept furthest when it produced fifty-five examples that it turned over to ordinary motorists for real-world evaluation. It soon became evident that the turbine's advantages of simplicity and light weight were negated by its inherent drawbacks: poor fuel economy, a pronounced lag when acceleration was needed, and a scorching exhaust. Turbines might have made some sense as power plants for heavy-duty trucks, where the ability to operate at relatively constant speeds for long distances partially offset the turbine's disadvantages, but continual improvements in diesel engines removed the motivation to pursue an expensive and risky strategy of trying to develop an altogether different kind of engine.

Another approach to direct rotary motion made its appearance in 1963, when Germany's NSU offered a small two-seat roadster powered by a Wankel engine. The Wankel, which was named after its inventor, Felix Wankel (1902–1988), represented an effort to avoid the reciprocal motions of pistons in a conventional engine by using a rotor shaped like a triangle with bulging sides that moved inside a housing shaped like an oval pinched in the middle. As the rotor revolved, it uncovered an intake port, compressed the air-fuel mixture, ignited it, and finally opened an exhaust port. The engine was lighter and more compact than a conventional piston engine, and the absence of reciprocating motions made it very smooth. On the debit side, it had poor fuel economy, and the rotor's tips were subject to rapid wear. Problems with the Wankel led to NSU's demise as an independent manufacturer, but the Japanese firm Toyo Kogyo continued to develop it for its Mazda line of cars.

## THE FOREIGN CAR INVASION

The vicissitudes of the Wankel-engined Mazda will be covered in the following chapter, which also will explore the remarkable growth of the Japanese industry as an automotive powerhouse. Japan did not become a significant force in the American market until the mid-1960s, but by this time the comfortable oligopolistic structure of the American automobile industry had already been challenged by cars imported from a resurgent Europe.

Imported cars had never comprised a significant part of the U.S. market, but in 1957 a threshold was crossed when for the first time sales of a foreign car, the Volkswagen, exceeded sales of some domestic makes. By 1965 the United States was importing a half-million foreign cars, with Volkswagen leading the pack. Although the Beetle could be criticized for its obsolescent design, cramped interior, leisurely acceleration, and spooky handling, its reliability and high-quality construction appealed to many motorists who had become disenchanted with the gaudy gas-guzzlers offered by the domestic manufacturers. Backing up VW's reputation for high quality was a dealer network that provided excellent post-sales service, as well as a quirky advertising campaign that appealed to motorists turned off by marketing fluff and planned obsolescence.

The export success of the Volkswagen also reflected the fact that although the car came up short in the horsepower department, it was not totally out of place in a country where the growing network of interstate highways required the ability to go long distances at relatively high speeds. Although the Volkswagens of the 1960s struggled to exceed 70 mph, once they got there they could comfortably cruise all day at that speed, a legacy of the VW's being designed for Germany's autobahns, many sections of which were unencumbered by speed limits.

The success of the Volkswagen stands as an instructive contrast with the inability of the British motor industry to maintain its position as the world's top exporter of automobiles. There were many reasons for the decline of the British automobile industry as an exporter and eventually even as a producer for its home market. One of the most important was that its products were designed for domestic consumption, and driving conditions in Britain were unlike those in the United States and many other export markets. While German cars had to prove their mettle on the high-speed autobahns of their place of origin, it was not until 1958 with the opening of Britain's first motorway, the M1, that British cars were subject to the requirements of high-speed cruising. To be sure, Britain had long produced Jaguars and Bentleys with superb all-around performance, but these constituted only a small slice of total production. The British car industry eventually was able to engineer moderately priced cars with good high-speed performance, but by this time the industry was confronting sharply declining sales and profits due to the poor quality of its products, inadequate investment, persistent labor strife, and destructive government policies.

Although it may not have been apparent at the time, while the British automobile industry was going into a slow, irreversible decline, Japan was beginning to assert itself as a significant force in the world automobile industry. This was an altogether unexpected development; in the years immediately following World War II, there seemed little likelihood that the defeated island nation would amount to much as far as automobile manufacture was concerned. As we have seen, its industry had shifted to truck production during the late 1930s, and there were probably only 80,000 cars in running condition when the war ended (Ruiz 1986, 16). Stimulated by America's need for trucks during the Korean war, the Japanese industry gradually came to life, but the cars it made during the next decade were thoroughly unimpressive. The small domestic market was fought over by eleven small- to medium-sized firms, and their products were for the most part copies of obsolete European designs. But by 1955 the Japanese industry was producing cars that at least were up-to-date in their basic specifications, with independent front suspensions, overhead valve engines, and modern (if not terribly attractive) bodywork.

In 1960 Japan had one car for every 240 people (Ruiz 1986, 19), and little emphasis was placed on exports. In 1957 its largest manufacturer, Toyota, took a stab at exporting cars to the United States, but its cars were so unsuited to U.S. conditions that the firm had to withdraw from the American market. In 1960 the entire Japanese industry succeeded in exporting only 7,000 cars, of which 942 went to the United States (Sedgwick 1983, 213). But, led by a surging domestic market, the Japanese car industry was able to substantially improve quantitatively and qualitatively, while a series of mergers resulted in fewer but stronger firms. In 1967, output topped the million mark for the first time, although the percentage of cars being exported was lower than that of European manufacturers.

Close observers of the Japanese automotive scene were impressed by the engineering and manufacturing capability embodied in such designs as the Datsun 510 and Toyota Corona sedans, the Honda S600 sports car, and the Wankel-powered Mazda coupes and sedans. In the years that followed, Japan moved from being a follower to a pacesetter, and the quality embodied in Japanese cars served to underscore the deficiencies of American and European cars.

## THE INTERSTATE HIGHWAY SYSTEM AND THE SUBURBAN DIASPORA

In 1939 nobody in the United States had the slightest inkling that European and then Japanese cars would begin to take a substantial share of the domestic market twenty years hence. Instead, the most powerful vision for the

General Motor's Futurama presented a future landscape dominated by high-speed, multi-lane highways. © Bettman/CORBIS.

year 1960 was the one presented by General Motors' exhibit at the 1939 New York World's Fair, "Futurama." By the time the fair ended, 5 million visitors had traveled in slowly moving seats while looking down upon a 36,000-square-foot diorama depicting the American landscape in the year 1960. The most notable feature was a grid of 12-lane highways, where futuristic cars glided by at high rates of speed, unencumbered by congestion, traffic lights, and all of the other dangers and annoyances of 1939-style motoring. It was a powerful image of a land remade by modern, high-speed highways, and even in the subsequent years when the nation was absorbed with fighting a total war, some members of the federal government were beginning to lay plans for a new highway system that, it was hoped, would turn this vision into reality.

As was noted in the previous chapter, the Federal Aid Highway Act of 1944 authorized an expenditure of $1.5 billion in matching funds over three years to build and improve the nation's highways after the conclusion of the war. The government took a far bolder step when President Dwight D. Eisenhower signed into law the Federal Aid Highway Act of 1956. The bill authorized spending $25 billion over a twelve-year period to pay for a National System of Interstate and Defense Highways, a planned network of more than 40,000 miles of multilane, limited access, toll-free roads. Congress also established the Highway Trust Fund to finance the enterprise. Using funds collected through taxes on gasoline, diesel fuel, tires, lubricants, and auto parts, the federal government would pay 90 percent of the Highway Trust Fund for construction and maintenance (Lewis 1997, 121), with the individual states responsible for the remaining 10 percent. The financing was described as "pay-as-you-go," and it had the advantage of avoiding interest charges that would have been incurred by issuing bonds. Paying for the initial stages of construction for the interstate highway system entailed an increase in the fuel tax from 2 cents to 3 cents per gallon, still very low by European standards of taxation.

The system's official title of "National System of Interstate and Defense Highways" reflected one ostensible reason for its construction. The Cold War era in which the system commenced was a time when the fear of an atomic attack was acute, and, according to some proponents, these highways would be invaluable for military mobilization and the evacuation of civilians from stricken areas. Some aspects of the system's construction, like mandating sufficient clearances for the transport of missiles on flatbed trucks, were attuned to military needs; in reality, few supporters took this aspect very seriously. Large-scale highway construction hardly needed military justification because it enjoyed support from an unusually broad coalition of interests: motorists, truckers, oil companies, construction workers' unions, public officials, and suppliers of building materials all stood to benefit from the system's construction and operation, and they all actively lobbied for its initiation.

The widespread support for the construction of the interstate highway system reflected an era when few questioned the benefits of automobile-based mobility. Even urban officials welcomed the penetration of the interstates into their cities, assuming that they would create jobs, improve commerce, and aid in slum clearance. Lewis Mumford, a longtime critic of the encroachment of the automobile into urban life, was distinctly in the minority when he characterized the interstate system as "an ill-conceived and preposterously unbalanced program" (quoted in Lewis 1997, 123). Over time, however, it became apparent that limited-access highways, many of them elevated over city streets, brought urban devastation in their wake. Large areas of homes and shops were demolished to make way for the new highways, and well-established neighborhoods were destroyed when the highways blocked access from one part of the neighborhood to another. And all the while, people living, shopping, and working in the vicinity of

Limited-access highways dominated sizeable portions of real estate throughout postwar America.

an urban freeway—many of them racial and ethnic minorities—had to contend with heightened levels of noise, air pollution, and visual blight.

Much of this devastation was of little or no consequence to the millions of people who used the new highways to commute from their homes to their places of employment. The interstate highway system was an integral part of one of the key social trends of the 1950s and 1960s: the relocation of a substantial part of the American population to the suburbs. This of course was not a new phenomenon; suburban living long antedated the coming of automobiles and interstates. Widespread ownership of cars and the expansion of highways represented the continuation of a long-term trend, for the growth of suburban living had always been closely tied to improvements in transportation. During the 1830s and 1840s, commuters used horse-drawn vehicles running on tracks or paved road surfaces to take them a few miles from their residences to places of work and then back home again. In the years following the Civil War, steam railroads and then electric trolleys made long-distance commuting possible, and many suburban housing developments sprung up in close proximity to railroad and trolley tracks. Although it allowed an escape from the perceived problems

Gigantic shopping centers like Minnesota's Mall of America depended on the Interstate Highway System to bring shoppers to them. Courtesy of Bordner Aerials, Inc.

of the city and made low-density, semirural living available to many, rail transport was inherently inflexible, confining housing to corridors closely adjacent to trolley or bus lines. In contrast, private automobiles could take commuters anywhere there was a road, and suburban developers put up tract homes in close conjunction with the building of new highways.

Migration to the suburbs closely paralleled the expansion of automobile ownership; between 1920 and 1930 the suburbs surrounding the ninety-six largest cities in the United States grew twice as rapidly as the cities themselves (Jackson 1985, 176). The trend was even more pronounced in the years following World War II; from 1950 to 1970, the number of people living in the suburbs more than doubled from 36 to 74 million, absorbing 83 percent of the nation's population growth (Jackson 1985, 283).

The distinguishing spatial characteristic of American suburbs has been its low population density. Single-family homes have been the dominant residential mode, a reflection of people's desires as well as government policies that encouraged their construction. The transportation needs of suburbia could not be met by public transit systems using buses and streetcars. Individualized public transit systems like dial-a-ride might serve a few suburban

Ownership of an automobile, such as this 1949 Ford Sedan, was an essential part of life in the suburbs. From the Collections of The Henry Ford.

residents; for the vast majority, however, the private automobile has been the only practical way of getting to work, running errands, and transporting children to after-school activities.

Automobile-based commuting also left its mark on domestic architecture. The front porch, once a kind of transitional zone between the private space of the home and the public space of the sidewalk and street, was not to be found in the great majority of houses built in the postwar era. In contrast, the garage, once kept out of sight in an area behind the house, now became a prominent feature of suburban residential architecture. In many instances, the doors of two- and even three-car garages were the most striking visual element of suburban homes.

## MOTELS, DRIVE-INS, AND FAST FOOD

Suburban idylls though their houses might be, people were not content to spend all of their free hours at home. Automobile travel was a well-established mode of recreation in the prewar years, and with the construction of the interstate system, road trips became easier and more extensive. One sign of the popularity of this mode of travel was the proliferation of roadside motels. Although the motel was an invention of the 1920s, it was

The first Holiday Inn began a tradition of promising "no surprises" for long-distance travelers. Picture of the 1952 Original Holiday Inn-Memphis appears courtesy of Inter-Continental Hotels Group.

not until the 1950s that the industry took the form that is so familiar today, the franchised chain offering uniform, "no surprises" lodging. The originator of the nationwide motel chain was Kemmons Wilson, who in 1952 built his first Holiday Inn in Memphis, Tennessee, and then expanded at a dizzying pace; by the end of the 1960s a new Holiday Inn was opening every three business days. Franchised motels closely followed the spread of the interstate highway system, and land adjacent to interchanges and off-ramps became prime areas for siting motel complexes.

In addition to seeking predictable lodging in nationwide motel chains, travelers on the interstates looked to franchised fast-food operations to satisfy their appetites. Limited-menu, fast-food restaurant chains emerged

A nighttime view of Stan's drive-in coffee shop where customers are served in their parked cars in Hollywood, California, on March 26, 1958. AP Photo.

in the 1920s, but the basic concept was taken to new heights by McDonald's. The first example, which was established by Richard and Maurice McDonald in San Bernardino, California, was able to serve a very large number of customers each day by restricting the menu to hamburgers, french fries, and drinks, and by stressing low prices, fast service, and uniform quality. A milkshake-machine salesman named Ray Kroc recognized the potential of the concept and bought the rights to franchise the McDonald brothers' operation. In 1955 the first franchised restaurant opened in Des Plaines, Illinois, followed shortly by two more in California. Five years later the chain encompassed 228 locations nationwide and was expanding rapidly. The concept soon attracted imitators that may have offered something other than hamburgers, but operated according to the same principles of fast service and uniform products. Not coincidentally, many of them could be found in close proximity to an interstate highway.

While the postwar automotive culture was altering the nature of travel, lodging, and dining, it also was changing the entertainment industry. One of the strongest trends of the 1950s and early 1960s was the growth of drive-in theatres. One thousand seven hundred drive-in theatres were serving the American movie-going public in 1950, and only four years later their number had grown to 4,200. But if drive-ins were a product of suburbanization and the expansion of the highway network, they were eventually destroyed by them, as the rising value of suburban land encouraged owners to seek more profitable uses for their property. After peaking at about 5,000 theatres in 1958, drive-ins began to decline in number, until by the end of the twentieth century they numbered only a few hundred, and many of these existed by serving as staging areas for swap meets and similar commercial ventures.

By the mid-1960s the top executives of the world's automobile industry had every reason to feel confident, even smug, about the future. The industry had enjoyed an amazing epoch of growth and development during the two decades following the conclusion of World War II. Automotive technology was advancing, car ownership continued to grow, and the built environment was being reshaped to suit the needs of an automobile-centered culture. Marketing disasters like the Edsel could be shrugged off, while triumphs like the Mustang added hundreds of millions of dollars to corporate coffers. But within a few years, confidence turned to profound anxiety as the seemingly impregnable position of the automobile came under attack from a number of quarters. The halcyon days were over, and as we shall see, the role of the automobile in modern society was being called into question as never before.

# 6

# Second Thoughts, 1965–1990

◆

By the mid-1960s the triumph of the automobile appeared to be complete. In the United States, cars were bigger, heavier, and more powerful than they ever had been, and their manufacturers were enjoying record levels of sales and profits. Across the Atlantic Ocean, mass motorization was taking hold in the industrialized countries of Western Europe. Britain, France, and Germany could claim more than two cars for every ten people, about half the U.S. average, but four to five times greater than the ratio of the early 1950s. Japan, which scarcely had an automobile industry a decade earlier, was churning out increasing numbers of cars for the domestic market and even exporting a fair number of vehicles to other countries.

Then it all seemed to come undone. Growing numbers of critics assailed the automobile for its noxious contribution to worsening air quality and for the tens of thousands of deaths it caused every year. Confidence in the viability of mass motorization was also severely shaken when an interruption in the supply of gasoline left many motorists wondering if they would be able to fill their tanks. The immediate crisis disappeared within a few months, but it did not put an end to ongoing concerns regarding the car's place in modern society. The years that followed were the scene of continual efforts to come to grips with widespread automobile ownership and the numerous problems it was causing.

## MAKING CARS SAFER

In 1965, 49,163 people died as a result of automobile accidents in the United States. On the one hand, this grim statistic represented continued improvement in automotive safety, as deaths per 100 million miles of motor vehicle travel had fallen steadily for decades, from 18.3 in 1923–1927, to 8.82 for 1947, to 5.36 in 1965 (Rae 1984, 138). Moreover, the 1961 ratio of road deaths to registered vehicles of 1 to 2,000 compared very favorably with Britain's ratio of 1 to 1,410, Germany's ratio of 1 to 430, and Italy's 1 to 375 (Rae 1984, 136). On the other hand, these accidental deaths represented a sum greater than U.S. losses in Vietnam during that year and all subsequent years. In any event, statistics that showed improved safety gave no comfort to the friends and family of the men, women, and children who had lost their lives or were seriously injured in traffic accidents.

As we have seen, Ford's trumpeting of the safety features of their 1956 model line met with tepid customer response, as usually could be expected with any new set of features. But the leaders of the American automobile industry were convinced that "safety doesn't sell," and they chose to interpret Ford's experience as a confirmation of that belief. Automobile executives did not have a callous disregard for the lives of their customers, but the bottom line came first, as it always had. For example, in the late 1920s Alfred Sloan resisted fitting General Motors' cars with safety glass because he felt that the additional costs would not be offset by greater sales, and GM's profits would suffer as a result (Rubenstein 2001, 213).

Such a view was to prove monumentally shortsighted. It is likely that concerns over automotive safety had been brewing in a public that was increasingly disenchanted with a domestic automobile industry that made styling its main selling point, and was ignoring practicality, safety, and quality. Under these circumstances, all that was needed was something that would focus growing concerns about automotive safety. That catalyst was provided in 1965 with the appearance of one of the most influential books of the second half of the twentieth century, *Unsafe at Any Speed: The Designed-In Dangers of the American Automobile.* The book had been written by Ralph Nader, a young lawyer who had been serving as an unpaid staff member of a Senate committee investigating automotive safety. The book led off with an indictment of the 1960–1964 Chevrolet Corvair, which was characterized as "a one-car accident." According to Nader, the Corvair's combination of rearward weight bias with a swing-axle rear suspension caused serious instability when cornering and a propensity to flip over when cornering limits were exceeded. The remainder of the book documented the "designed-in dangers" of other automobiles: power brake systems that failed suddenly and completely; steering column–mounted indicators for automatic transmissions that led drivers to shift into reverse when they intended to shift into low, and vice versa; dashboard reflections that impeded vision through the windshield; steering wheels that impaled drivers in the event of a crash; and control knobs that caused serious injuries in low-speed crashes. Even more disturbing, the book also provided ample evidence that the industry had been indifferent or even hostile to the idea of making safer cars and to safety-related regulations that might lead to government oversight of the automobile industry.

At this point the industry's leaders had been completely successful in fending off federal government involvement in the design of automobiles, but their fears about an expanded role for the government were about to be realized. In 1964 the General Services Administration (GSA) began working on the establishment of safety standards for cars purchased by the federal government. Since the government bought 36,000 vehicles annually at that time, it was in a good position to influence automobile design. The standards were published the following year, but they required no changes of any real substance, not a surprising outcome given the GSA's unwillingness to lock horns with one of the nation's most powerful industries.

At this point, the efforts of Nader and a few others to spotlight the issue of automotive safety seemed to be arousing little interest. Then, General Motors, which at the time produced more than half the vehicles sold in the United States, made a blunder of massive proportions. Annoyed by the accusations made in *Unsafe at Any Speed*, the chief counsel of General Motors hired a private detective to look into Nader's personal life in the hope of finding something unseemly. The investigation uncovered nothing of interest, and the public was outraged when GM's snooping was revealed. Nader sued GM for invasion of privacy, and eventually settled out of court for $425,000, which he used to fund a public interest advocacy organization.

General Motor's investigation of Ralph Nader had been only the latest manifestation of the automobile industry's arrogance. In 1966 Congress reined in the industry by passing the National Traffic and Motor Vehicle Safety Act, which for the first time empowered the federal government to set safety-related standards and to order recalls of vehicles with defects that affected their safety. The act also created a new regulatory agency under the Department of Transportation, the National Highway Traffic and Safety Administration (NHTSA, often pronounced "Nitza"), which came into being the following year. One of NHTSA's first tasks was to draft Federal Motor Vehicle Safety Standards to set performance criteria for brakes, lights, and tires. These standards also attempted to mitigate the effects of the "second collision" between a car's occupants and its interior by mandating the redesign of potential sources of injury, such as projecting control knobs. The agency also required the accumulation of data by manufacturers to aid in safety research.

One of the most contentious issues of automotive safety centered on how to keep drivers and passengers from being flung about a car's interior during a collision or, even worse, being ejected from the car. The first step involved a requirement that all cars sold from the 1968 model year onward had to be equipped with seat belts. Accident statistics made it clear that unrestrained drivers and passengers were far more likely to suffer death or injury than those using seat belts, but the problem was getting people to use them. In 1974 NHTSA tried to accomplish this by requiring all new cars to be fitted with seat belt interlocks that prevented a car from being started if the occupants' seat belts were not fastened. This rule met with such hostility from car buyers that it was soon rescinded by a special act of Congress.

Convinced that the public would not voluntarily use seat belts in sufficient numbers, safety officials mandated the equipping of cars with "passive restraints" that required no action on the part of drivers and passengers. Some manufacturers met this requirement, which began in 1990, by fitting their cars with automatic belts that used electric motors to move them into place. These were widely viewed as an annoyance, and they were abandoned after a few years in favor of airbags that automatically inflated in the event of an accident. Airbags had been offered as an option on some Oldsmobiles, Buicks, and Cadillacs cars beginning in 1973, but only about 10,000 customers had been willing to spend the $181 to $300 required, well short of the 50,000 anticipated. General Motors dropped them in 1976 after claiming substantial financial losses as a result of the program. But customers' attitudes appeared to be changing, and airbags met with a more positive reception in the early 1990s, especially when the alternative was an annoying motorized harness. Fifty-six percent of the cars sold in the United States were equipped with a driver's side airbag in 1992, and three years later they were almost universal on new cars, while only 5 percent lacked a passenger-side airbag. At the time, this seemed like a major safety advance, but as we shall see in the next chapter, airbags turned out to be a very mixed blessing.

Experts still disagree on how to weight the various factors that affect automobile safety in the aggregate. Such disparate influences as the age distribution of the driving population, the quality of emergency room services, the state of the economy, highway design, and the extent of drug and alcohol use all affect accident statistics and fatality rates. Nonetheless, it does seem evident that improvements in the design of automobiles have made a substantial contribution to the diminishing accident fatality rates. Automotive

**Table 6.1**
**Accident Fatalities, 1980–1994**

| Year | 1980 | 1985 | 1990 | 1994 |
|---|---|---|---|---|
| Total fatalities (in thousands) | 53.2 | 45.9 | 46.8 | 42.5 |
| Fatalities per 100,000 registered vehicles | 34.8 | 26.4 | 24.2 | 21.2 |
| Fatalities per 100 million vehicle miles | 3.3 | 2.5 | 2.1 | 1.7 |

Adapted from U.S. Census Bureau, *Statistical Abstract of the United States*, 2001, Table 1092.

engineers learned how to improve "passive" safety by designing cars with crumple zones to absorb the energy of a collision, while at the same time "active" safety was improved by equipping cars with better brakes and suspensions. In sum, automotive safety has shown marked improvement over the years, as can be seen in Table 6.1.

Driving still has its hazards, but there can be little doubt that from the 1960s onward, engineering advances made major contributions to automotive safety by decreasing the number of accidents and minimizing their consequences when they do occur.

## CLEARING THE AIR

In the 1940s the residents of the Los Angeles area began to notice that on many days the sky had taken on a brownish-gray cast, an acrid smell permeated the air, and many residents were experiencing some difficulty in breathing. The phenomenon was labeled "smog," and it was to become a growing problem in Southern California and other parts of the country.

Although the word "smog" was derived from a blending of "smoke" and "fog," it really is something quite different. Smog is the product of reactions involving a number of chemical compounds that take place under the influence of sunlight. For this reason it is sometimes referred to as "photochemical smog." The basic constituents of photochemical smog are formed when the combustion of a carbon-based fuel results in the production of unburned hydrocarbons and oxides of nitrogen. The ultraviolet radiation of sunlight breaks the oxides of nitrogen into nitrogen dioxide ($NO_2$), some of which reacts with the unburned hydrocarbons to form ozone ($O_3$), and peroxyacyl nitrates (PANs). In addition, the combustion of fossil fuels produces particulates (very small particles of carbon), carbon

Los Angeles City Hall enshrouded in smog. Courtesy of the Library of Congress.

monoxide (CO), carbon dioxide ($CO_2$), and sulfur dioxide ($SO_2$). These compounds can be very problematic, but strictly speaking they are not components of photochemical smog.

When smog was first recognized as a serious problem there was some question regarding its origins, but in the early 1950s Arie Haagen-Smit at the California Institute of Technology proved that automobile exhaust was one of the chief culprits. This conclusion was bitterly disputed by the automobile industry, which asserted that well-tuned cars in good repair did not contribute significantly to air pollution. The obstructive attitude of the industry could not be maintained as severe smog episodes in Southern California became more frequent, and other parts of the country began to suffer from heightened levels of air pollution. If better air quality was to be achieved, automotive emissions had to be addressed; in the 1970s it was estimated that cars and trucks were responsible for 75 percent of carbon monoxide, 35 percent of unburned hydrocarbons, and 29 percent of the nitrogen oxides released into the nation's atmosphere. Although other sources were collectively responsible for the majority of smog-causing emissions, there was no question that motor vehicles were important contributors to air pollution.

California, the state most severely affected by air pollution, began to take the first steps to control automotive emissions. In 1961 it mandated the fitting of positive crankcase ventilation (PCV) valves on all new cars sold from 1963 onward. These reduced the hydrocarbon emissions that blew past piston rings into engine crankcases and then were vented into the air. In 1966 the California legislature took a significant step when it established standards that limited emissions of carbon monoxide, oxides of nitrogen, and lead compounds for vehicles sold in the state.

By this time the federal government also had become involved in the control of automobile emissions. The 1965 Motor Vehicle and Air Pollution Act limited hydrocarbon and carbon monoxide emissions from new cars, although it allowed higher levels than the ones mandated in California. In 1970 Congress passed a more ambitious piece of legislation, which was entitled the Clean Air Act. This stipulated a reduction by 1975–1976 of carbon monoxide emissions to 0.41 grams per mile, hydrocarbons to 3.4 grams per mile, and 0.40 grams of nitrogen oxides per mile (average motor vehicle emissions at that time were 46, 4.7, and 6.0 grams per mile for carbon monoxide, hydrocarbons, and nitrogen oxides, respectively). The act gave a newly created agency, the Environmental Protection Agency (EPA), the power to enforce these standards. Rather predictably, the automobile manufacturers claimed that they could not meet this timetable, and the deadline was extended by two years, even though Honda had been building cars that already were in compliance.

In fairness to the domestic automobile manufacturers, it has to be said that reducing automotive emissions is a difficult task. A conventional Otto-cycle engine allows very little time to combust the air-fuel mixture, and some residue of unburned hydrocarbons is inevitable. The combustion process can be improved by raising compression ratios, but the resulting higher temperature in the combustion chamber results in the formation of more oxides of nitrogen. Consequently, efforts to improve emissions performance often entail tradeoffs; the reduction of hydrocarbon emissions increases emissions of oxides of nitrogen, while the reduction of oxides of nitrogen causes more hydrocarbon emissions. The situation is further complicated by the need to take account of the engine's constantly changing rotational speeds, its fluctuating operating temperature, and the unfortunate tendency of many motorists to ignore necessary maintenance.

The difficulty of reducing emissions was all too evident when the first generation of emissions-controlled vehicles were put on the road. In their efforts to produce cleaner cars, manufacturers resorted to a number of stop-gap technologies like air pumps to promote combustion of unburned fuel, lowered compression ratios, and very lean air-fuel mixtures. The result was diminished performance, higher fuel consumption, and "drivability" problems like surging, stalling, and dieseling (running on after the ignition had been switched off). Many observers predicted that the days of the conventional internal combustion engine were numbered, and that it would have to be replaced by something altogether different; even a rebirth of steam-powered cars was deemed a possibility.

As it turned out, the Otto-cycle engine demonstrated remarkable staying power. Although its basic principles go back to the nineteenth century, the conventional internal combustion engine has been able to maintain its dominance as an automotive power plant through the incorporation of technologies that emerged in the late twentieth century. Several of these were devoted to improving combustion performance. If an engine is to effectively and efficiently burn all the fuel fed into its combustion chamber, it has to keep the ratio of air to fuel within a narrow range, which requires adjusting flow of air and fuel in accordance with altitude changes and other external factors. Conventional carburetors are not terribly effective in doing this. A much better proposition is fuel injection, which supplies a precisely metered quantity of atomized fuel into the combustion chamber.

Fuel injection is not a new technology; Bosch in Germany manufactured a system in 1912, and it had been an essential component of diesel engines since the 1920s. In the 1930s fuel injection was used for high-performance aircraft, which flew at a wide range of altitudes and were put through severe aerial maneuvers. Fuel injection was first employed for automobiles other than diesels when Daimler-Benz drew upon its prior experience with aircraft engines to employ it in the Mercedes 300SLR sports racing car and some of its passenger cars in the early 1950s. Fuel injection was used by a number of other cars in the years that followed, one of the most noteworthy being the 1957 Corvette, which was the first American production car to claim one horsepower for each cubic inch of displacement.

These early systems worked effectively, but they were expensive to manufacture and required skilled maintenance. With tightened emissions regulations, manufacturers realized that the carburetor's days were numbered and fuel injection had to be fitted as standard equipment. Volume production drove down costs, and electronic controls accurately and reliably adjusted fuel flow to take into account variables like throttle position and engine temperature.

The application of electronic technologies to automobile engines also allowed significant improvements in automotive ignition systems. Conventional distributors could only effect rather gross changes in timing the delivery of a spark to an engine's combustion chamber. Much more accurate ignition timing was made possible by the use of computerized controls that fired an engine's spark plugs in accordance with the vehicle's speed, the gear engaged, and the octane number of the gasoline being used. Together, these new engine management technologies greatly improved emissions performance without creating all of the problems that attended such efforts in the 1970s.

Emissions also were improved through the use of reformulated gasoline. In part, this was done through the removal of substances like lead and sulfur that made their own noxious contributions to the polluted skies. In addition, hydrocarbon and nitrogen oxide emissions were reduced by removing the more volatile olefins and aromatics from gasoline. Emissions performance was further improved by altering gasoline blends in accordance with local conditions such as elevation and climate.

All of these changes resulted in significantly cleaner exhausts. However, the technological centerpiece of the effort to clean up car emissions was the catalytic converter. These devices, which were first mandated by the state of California for all 1975 model cars, are fitted to the exhaust system in front of the muffler to clean up tailpipe emissions. The first of these were known as two-way converters because they supported a catalyzed process of oxidation that converted unburned hydrocarbons and carbon monoxide into carbon dioxide and water. These were soon followed by three-way converters that also supported a reduction process that turned oxides of nitrogen back into free nitrogen and oxygen.

The effective deployment of catalytic converters required the use of many of the antipollution processes and technologies described above. Lead had to be removed from gasoline not just because it was a toxic substance, but also because it seriously damaged catalytic converters. Catalytic converters also depend on computerized engine management systems to maintain precise air-fuel mixtures, thereby preventing excessive rates of either oxidation or reduction.

## MORE MILES TO THE GALLON

For most of the twentieth century, the cars made in North America diverged substantially from the products of European and Japanese manufacturers. The most significant difference was size; American cars were considerably larger and heavier, and they needed much larger engines to pull all that weight. By the mid-1960s the differences were striking. On average, American cars were at least 1,500 pounds heavier, and the V-8 engines commonly found under their hoods had more than twice the displacement of the four-cylinder engines commonly found in European and Japanese cars. As a result, American cars consumed much more fuel than automobiles built elsewhere; in 1974 American cars averaged 13.2 mpg, while the cars imported from abroad averaged 22.2 mpg (Flink 1988, 388).

The date is significant, for it marks the time that the manufacturers of American automobiles and the people who drove them suddenly came to the realization that the profligate use of gasoline could incur steep costs. In 1973–1974, gasoline supplies, which had been taken for granted for so long, were suddenly constricted, and motorists had to wait in long lines in order to fill their tanks. There still was plenty of petroleum in the ground, but unbeknownst to most Americans, their country had become increasingly dependent of oil from abroad. The United States was importing 27 percent of the oil it used (Flink 1988, 389), and a significant portion of that supply came from a particularly volatile part of the world, the Middle East. In 1973 the region boiled over when on October 6, Egypt and Syria attacked Israel in what became known as the Yom Kippur War. By December of that year, the Middle Eastern members of the Organization of Petroleum Exporting Countries (OPEC) cut off the supply of oil to the United States and the Netherlands as punishment for their support of Israel. The exporters raised the price of crude oil for those countries that were not subject to the embargo, and other oil-producing countries followed their lead. At this point, retail gasoline prices in the United States were subject to price controls imposed by the Nixon administration, so refiners and distributors chose to supply less gasoline when prices at the pump did not reflect what they were paying for crude oil. To make matters worse, many motorists reacted to the shortages by filling their tanks even when they were still more than half full, further jamming the lines at gas stations. In sum, shortages of gasoline coupled with panic over its availability led to a crisis situation that severely shook the previously taken-for-granted notion that cheap gasoline would always be available.

Gasoline availability eventually returned to normal, albeit at higher prices, but long gasoline lines returned in 1979 when petroleum supplies were disrupted by the Islamic revolution that toppled the Shah of Iran. Once again, the availability of gasoline in the United States had been severely affected by events transpiring halfway around the world.

Many motorists responded to shortages and rising prices by switching to smaller cars, many of them of foreign origin. The first energy crisis also was followed by more involvement of the federal government in the design of automobiles; to government regulation of safety and emissions were now added government-mandated fuel economy standards. In 1975 the Energy Policy and Conservation Act stipulated that the cars produced by

American manufacturers had to meet a corporate average fuel economy (CAFE) standard of 18 mpg in 1978, and 27.5 mpg by 1985. Failure to achieve these averages would result in fines of $5 for every tenth of a mile below the standard, multiplied by the manufacturer's total production for that year. The government further tightened the screws in 1978 when it established a "gas guzzler" tax to be levied on cars with high fuel consumption. In a further move to conserve gasoline, the government mandated a nationwide speed limit of 55 mph that went into effect on January 1, 1974. It was partially lifted in 1987, when individual states were allowed to increase speed limits to 65 mph on rural highways.

## REDESIGNING THE AMERICAN AUTOMOBILE

Twenty years before the energy crises of the 1970s, a few American automobile manufacturers produced car models that were smaller than the typical Detroit product (but larger than most European cars). The lukewarm reception accorded to cars like the Hudson Jet, Kaiser Henry J, and Willys Aero reinforced the industry's conviction that Americans were not interested in small, fuel-efficient cars with limited power and interior space. This apparent lack of demand for smaller cars was welcomed by the major domestic manufacturers, for they had long shared the opinion of Henry Ford II that "mini cars mean mini profits." As a result, the manufacturers and the people who bought their products shrugged off the first fuel crisis, and at the end of the 1970s, 80 percent of American cars still were powered by V-8 engines (Rae 1984, 142). But rising gasoline prices expanded the market for gas-saving cars, and Japanese and European manufacturers were well positioned to supply that market. Between 1978 and 1980, domestic manufacturers sold 2.5 million fewer cars while sales of foreign imports, 80 percent of which were Japanese, increased by a half million (Rae 1984, 156).

Datsuns, Toyotas, and Hondas appealed to increasing numbers of motorists because they saved gasoline and had gained a solid reputation for reliability, as evidenced by the fact that of the 6.5 million cars recalled to remedy defects in 1979, only 3.8 percent were of Japanese origin, even though they constituted 14 percent of the cars sold in the United States in that year. The demand for energy-efficient automobiles stimulated the sales of Japanese cars, but it was not the only reason that growing numbers of automobile buyers were abandoning domestic products.

American automobile manufacturers were being squeezed from two sides; a fair portion of their market was now being taken by foreign manufacturers, while government CAFE standards were limiting their ability to sell their favored product, the powerful, oversized land yacht. The industry's response, which came to be known as "downsizing," entailed nothing less than the large-scale reengineering of domestically produced cars.

Even before events of the 1970s forced widespread changes in the design of American cars, competition from foreign manufacturers had impelled Detroit to offer its own line of fuel-efficient cars. The surest way to increase a car's fuel economy is to reduce its weight, so just as they had done in the late 1950s, domestic manufacturers introduced several models of smaller, lighter cars. The first of the breed were the Ford Pinto and Chevrolet Vega, which were introduced for the 1971 model year. Neither did much to convince people that domestic manufacturers had the ability to produce world-class small cars. Both were styled to give a sporty appearance at the expense of interior space. Making matters worse, the Vega was plagued with numerous reliability problems, while the Pinto had a reputation, perhaps unearned, of bursting into flames when hit from behind.

In a more serious effort to design a new generation of space-efficient small cars, General Motors turned to front-wheel drive, which it had already employed on its "full-size" Oldsmobile Toronado. Its initial front-wheel-drive offerings, the 1980 Chevrolet Citation, Pontiac Phoenix, Oldsmobile Omega, and Buick Skylark, left a lot to be desired, as they gave ample evidence of having been rushed into production. At least General Motors had a line of cars appropriate for the times. In contrast, Chrysler's offerings, which had been restyled a few years earlier, were the biggest the company had ever produced. The results were predictable; large numbers of Plymouth Gran Furys, Dodge Royal Monacos, and Chrysler New Yorkers piled up in dealers' lots as the company's financial situation became increasingly desperate. With losses of $1 billion in 1979, Chrysler was rapidly heading toward bankruptcy, and if it went down, it would take 97,000 production workers' jobs with it. To prevent this from happening, the federal government agreed to put up $1.5 billion in loan guarantees to keep the firm afloat until it came out with a new line of fuel-efficient, front-wheel-drive cars. In return for the guarantees, Chrysler's managers and workers had to tighten their belts. The wage, benefits, and work rule concessions made by Chrysler's workers were slightly offset by the government's requirement that a union representative had to be given a place on the board of directors. In due course it was occupied by Douglas Fraser, the president of the United Auto Workers Union. This was a fairly common arrangement for European firms, but it represented a first for the American automobile industry.

The Chrysler "bailout" allowed the firm to borrow money from financial institutions to tide it over until the arrival of its new generation of front-wheel-drive vehicles, the K-cars. Marketed as the Plymouth Reliant and the Dodge Aries, the cars were commercial successes, and equally important they became the basis for a whole new category of vehicle, the minivan. Although there had been a few precursors to the Chrysler product such as the Volkswagen Microbus and the Chevrolet Corvair Greenbrier, these vehicles had a number of shortcomings and they sold in modest numbers. Chrysler's minivans, which were first offered to the public in early 1984, appealed to motorists because they had a great deal of interior volume for people and things, and although they couldn't be described as nimble, they didn't have the trucklike driving characteristics of full-size vans. Most importantly for Chrysler, they were the right product at the right time. Twenty years earlier the baby boom generation had made the Mustang a runaway success. Now as they approached middle age, the boomers needed something more practical to transport children to soccer games and to haul bags of fertilizer from the nursery. The minivan was the perfect vehicle for these duties, and more than any other product it led the Chrysler Corporation back to profitability.

## WAGES AND WORKING CONDITIONS IN THE AUTOMOBILE INDUSTRY

Innovative products like minivans and more fuel-efficient cars helped to revive the American automobile industry, but its output of 6.8 million cars and small trucks in 1989 was considerably lower than the 9.3 million vehicles produced in 1965 or the 9.7 million produced in 1973. The employment statistics were even worse because the industry shed many jobs during this period by automating and outsourcing (buying components from outside manufacturers). To take one example, Ford made about as many vehicles in 1988 as it did in 1978, but it was able to do so with half as many employees. Those workers who were able to keep their jobs did well financially, as wages and benefits continued to be about double those earned in the manufacturing sector as a whole, but these gains did not come easily.

In fighting for better wages and benefits for its members, the United Automobile Workers Union continued to use tactics that had been effective for many years. It selected General Motors, Ford, Chrysler, or American Motors as the targets of its triennial bargaining efforts over wages, benefits, and hours. Once a settlement was reached, it was used as the basis of the UAW's contracts with the other manufacturers. General Motors was rarely selected as a strike target because it was the industry's biggest and most powerful firm, but some of the industry's longest strikes ensued when this occurred. In 1945–1946 production was halted when the union went on strike for 113 days. Another protracted action took place in 1970 when 340,000 GM workers walked off the job. As the strike took hold, tens of thousands more workers in firms that supplied parts to GM were also idled. GM lost $90 million a day in sales, suppliers lost $40 million each day, and local governments lost $20 million a day due to reduced tax revenues (Serrin 1974, 186–187). After sixty-seven days, GM and the UAW reached an agreement. UAW workers gained increased wages and benefits, along with a somewhat accelerated timetable for early retirement, but GM suffered no harm; it went on to earn its customary 20 percent on investment during the three following years when the contract was in force (Serrin 1974, 312).

As had been the case with previous negotiations, neither GM nor the union showed much interest in quality-of-working-life issues, and the 1970 settlement did nothing to alleviate the repetitive, mind-numbing, exhausting labor of the assembly line. By this time factory automation had made substantial progress, and optimists hoped that robot welders and the like would eventually replace most assembly-line labor. The General Motors plant at Lordstown, Ohio, represented just such a state-of-the-art plant. Its high level of automation allowed it to operate at a line speed of 104 Chevrolet Vegas per hour instead of the 55 to 75 cars per hour typical of the industry. But hopes that Lordstown represented a factory of the future were dashed in 1972 when its 8,000 workers went out on strike to protest the excessive line speed and the firing of 300 workers who had protested the line's pace. Lordstown became a national symbol of the "blue-collar blues," which seemed especially evident among younger workers, many of whom had served in Vietnam.

Although the speed at which the Lordstown line operated made it an extreme case of worker unhappiness, other automobile factories still had to cope with an alienated workforce whose disaffection could be gauged by daily absenteeism rates of 5 to 7 percent. The frustration and hostility of assembly-line workers became well known to potential car buyers, who were admonished to avoid buying a car that had been made on a Monday or a Friday, the days when absenteeism was the greatest and worker motivation was the lowest. Some customers even reported that their new cars appeared to have been deliberately sabotaged at the factory.

The most radical response to the human problems caused by assembly-line work was Volvo's innovative assembly works at Kalmar, Sweden, which did away with the assembly line altogether. Instead of doing the same operations over and over, workers at Kalmar performed a number of tasks that could be changed from time to time. Workers belonged to teams of 15–20 workers who collectively decided how to distribute the work within the team. As an experiment, Kalmar was a modest success. It increased manufacturing costs by about 10 percent, a significant financial penalty in an industry that is always trying to make cars as cheaply as possible. Moreover, Kalmar was a small operation in which 600 workers assembled 30,000 vehicles a year, so extrapolating from it would probably entail additional costs. The work performed was limited to final assembly; the factory did not do metal stamping, foundry work, or body welding—some of the most arduous jobs in automobile manufacture. Finally, the apparent improvements in the work environment apparently were not reflected in a more dedicated workforce; the absenteeism rate was quite high at 15 percent, although it may have been inflated by Sweden's very generous sick-leave policy.

A more lasting impact on workers' experiences in the automobile industry came from practices developed in the Japanese automobile industry, where a higher level of job satisfaction seemed to prevail. Strikes were certainly not unknown, but in general, labor-management relations seemed not to be as acrimonious as those in the United States and Europe. In part, this was due to the absence of the clear-cut distinction between workers and management that was evident in the U.S. automobile industry. Japanese managers worked in open cubicles instead of private offices and shared dining facilities with workers, and the visual distinction between blue- and white-collar workers was eliminated by having everybody wear the same company uniforms.

Japan also pioneered the use of quality-control circles that involved workers in at least some shop-floor decisions. The empowerment of line workers was highlighted by a practice that had considerable practical and symbolic importance: they had the right and the duty to stop the line in the event of a problem with components or the assembly process. By contrast, in American or European factories the ability to halt production was the exclusive right of shop-floor managers, who were convinced that workers would not behave responsibly if they had this power.

The benefits of Japanese management practices were apparent when General Motors and Toyota entered into a joint agreement to form a firm known as New United Motors, Incorporated (NUMMI), that would use a former Chevrolet plant in Fremont, California, to assemble Toyota Corollas to be sold as Chevrolet Novas. The factory had been plagued by the usual ills of domestic automobile factories—absenteeism, on-the-job drug and alcohol use, and a high level of tension between workers and managers. The situation began to turn around when production of the Nova commenced in 1985. Under Toyota's management system, worker morale was much better, productivity was up, absenteeism dropped from 20 percent to

2 percent, and quality was noticeably better. A fair portion of this overall improvement may be attributed to the fact that although 80 percent of the workforce was drawn from the ranks of former GM workers, very few former GM managers were taken on by the new venture.

## JAPAN MOVES INTO FIRST PLACE

The joint General Motors–Fremont operation was part of a much larger trend, Japan's ascension to the top rank among the world's automobile manufacturers. Japan's automobile industry had experienced explosive growth in the 1960s, and by 1968 Japanese manufacturers were producing more than 2 million cars, putting the country in fourth place among the world's car-manufacturing countries, ahead of France and Italy. In 1971 Japan became the world's second largest producer of cars, and three years later it passed Germany to be the world's largest exporter.

Some of the industry's strength was derived from the profits it earned from sales in a rapidly expanding domestic market that was largely closed to cars made outside Japan due to tariffs and other restrictions imposed by the Japanese government. The government also tried to create greater economies of scale by merging smaller firms to create larger ones. Government bureaucrats also attempted to prevent Honda, the world's largest motorcycle manufacturer, from getting into automobile production. These efforts met with only limited success. Honda was not deterred from entering the car industry, and of the seven mergers proposed by the government, only three actually occurred when Nissan acquired Prince, and Toyota took over Hino and Daihatsu.

In the final analysis, the remarkable success of the Japanese automobile industry derived from the cars it made. Customers worldwide recognized the quality, reliability, and all-around excellence of Japanese products like the Honda Accord, the Datsun 240Z, and the Toyota Celica, and they voted their approval with their checkbooks.

The formidable competition offered by Japan produced a political backlash, as domestic manufacturers in a number of countries were able to convince their governments to limit the influx of Japanese cars and small trucks. In the 1980s France restricted Japanese imports to 3 percent of the market annually, Britain held them to 10 percent, and Italy allowed a mere 2,000 Japanese vehicles to enter the country each year. In the United States, the Nixon administration tried to stem the growing sales of Japanese cars by slapping a 10 percent surtax on Japanese imports, and in 1981 the Reagan administration put aside its commitment to free trade to negotiate "voluntary" export restrictions known as "orderly marketing arrangements." These limited the number of Japanese imports to 1.68 million per year, rising slightly to 1.85 million cars two years later. In subsequent years the Japanese government unilaterally raised the quota to 2.3 million.

Although these allocations were intended to take some of the pressure off American automobile manufacturers, in the long run they proved to be of great benefit to the Japanese automobile industry. Restricted supplies resulted in higher prices and more profit for Japanese (and American) producers. Also, since the quotas were expressed in the number of vehicles exported, Japanese manufacturers had a strong incentive to export more expensive cars with higher profit margins. They therefore created new model lines like Lexus, Acura, and Infiniti, which went on to pose a powerful challenge to the established manufacturers of luxury cars. Finally, import restrictions motivated Japanese manufacturers to build cars in the market that they were sold. The first to do so was Honda, which was already assembling motorcycles in Marysville, Ohio. In 1982 Honda began producing Accords in the expanded facility, and a year later it was followed by Nissan, which began to produce small pickup trucks in Smyrna, Tennessee, followed by the manufacture of Sentra cars in 1985. Within a few years, all of the major Japanese manufacturers had established U.S. operations—Toyota in Georgetown, Kentucky; Subaru and Isuzu in Lafayette, Indiana; Mitsubishi in Normal, Illinois; and Mazda in Flat Rock, Michigan. Significantly, only the last two of these had a workforce that chose to be affiliated with the United Auto Workers Union. It is also noteworthy that the three unionized plants (NUMMI being the other one) started as joint ventures between Japanese and American firms.

## THE MOTORIZATION OF THE SOVIET UNION

While Japan was becoming a dominant force in the global automobile industry, the Soviet Union was making a belated effort to make more cars available for its citizens. In the late 1960s, a mere one out of 200 people owned a car and even those with the necessary funds had to wait two years for delivery. The Communist leadership recognized that private automobile ownership was the paramount symbol of an improved standard of living, and that it had to expand automobile ownership as a sign of the superiority of socialism. This was a difficult task because the Soviet planned economy had a poor record of promoting innovation, and it never had been much good at meeting the needs of consumers. To break out of this impasse, the Soviet government used the same strategy it had employed in the 1930s:

it enlisted foreign companies to take the lead in expanding and modernizing its automobile industry. Fiat was recruited to build a huge automobile factory at a site by the Volga River, which was accompanied by the construction of a new town named "Togliattigrad" in honor of the longtime head of the Italian Communist Party. At about the same time Renault was called in to modernize the decrepit Moskvitch factory in Moscow, and it also provided major assistance in building a giant truck factory on the Kama River in the 1970s.

These projects greatly increased vehicle production in the Soviet Union. In 1973 the industry produced 550,000 copies of the Fiat 124 sedan, which were known as "Zhigulis" in the USSR and "Ladas" when they were sold abroad (at very attractive prices). Three hundred thousand of these were destined for export, and a similar number went to government ministries and for use as taxis, so ordinary Soviet citizens still had to endure long waits before they got their cars Even so, the Soviet Union now occupied ninth place among the world's automobile-producing nations, and automobile ownership would grow significantly in the years that followed.

A similar pattern could be observed in other East Bloc countries. Assistance from Fiat allowed the Polish automobile industry to manufacture more than 100,000 copies of the Fiat 125 in 1973, and in the same year the Romanian automobile industry produced more than 50,000 examples of the Renault 12. Yugoslavia, which had managed to remain free of Soviet domination, also looked to Italy to energize its automobile industry. Fiat's assistance allowed the Zastava factory to produce upwards of 60,000 cars annually during the 1970s. Some of them were even exported to the United States under the brand name of "Yugo," which eventually rivaled the Edsel as fodder for automotive humor.

## THE MELTDOWN OF THE BRITISH CAR INDUSTRY

While the Soviet Union was rapidly expanding its automobile industry and Japan was challenging the United States to be the world's largest producer of automobiles, the British automobile industry was beginning to unravel. The world's second largest producer of automobiles in the years immediately following World War II, Britain had fallen to sixth place in 1975 and seventh place in 1980. Among its manufacturers, Ford of Britain showed consistent profits, but the other two firms with American ties did much less well. Vauxhall, General Motors' British operation, remained a marginal player. Chrysler gained majority control over Rootes 1967, but sold it to Peugeot in 1978 after years of poor financial returns.

Especially sad was the fate of the only major segment of the industry that had no ties to American firms. The postwar consolidation of the British-owned automobile industry began in 1952 when the merger of Austin and Morris produced the British Motor Corporation (BMC), which then merged with Jaguar in 1966 to form British Motor Holdings (BMH). The last major consolidation followed in 1968 when BMH merged with Leyland Motors, a truck and bus manufacturer that also produced Triumph and Rover cars. The resulting firm was christened British Leyland. The merger failed to solve the indigenous car industry's problems, and from 1975 to 1988 the British government poured £3 billion into British Leyland in order to prevent the loss of tens of thousands of jobs (Church 1995, 111). And in an effort to upgrade the quality of its cars, in 1981 British Leyland began producing Honda cars to be sold as Triumphs. None of this activity stopped the erosion of market share, which had skidded to 15.5 percent of the home market and was accompanied by a loss of £892 million in 1986 (Wood 1988, 244).

The reasons for the decline of the British automobile industry were numerous. Conflicts between labor and management have often been singled out as the key source of the industry's troubles, and there is no question that they constituted a serious problem; in 1965 more than 8 million hours were lost as a result of strikes (Wood 1988, 135), and some subsequent years were even worse. But management also has to take its share of the blame. British Motor Corporation and its successor firms never succeeded in rationalizing their model lines and their manufacturing facilities. This resulted in high production costs and divided marketing efforts. Management shortcomings were also evident in the firm's underdeveloped accounting techniques. It was noted in the previous chapter that its signature product, the Mini, earned scanty profits at best, and the company could not even determine which models made money and which produced losses. The British government also had a hand in the industry's eventual collapse. New production facilities were required to be located in areas of high unemployment, but these were far from the major sources of supply and skilled labor, resulting in inefficiencies and increased costs.

The combination of an excessively diverse line of car models, aging plants, the suboptimal location of new ones, poor management, and labor strife had unfortunate consequences for productivity. In 1974 each British Leyland employee accounted for £2,129 ($5,110) in value added; comparable figures were £4,767 ($11,441) for Volkswagen and £8,600 ($20,640) for General Motors' operations in the United States. To make matters worse, the taxation and monetary policies of both Labour and Conservative governments subjected the industry to an endless stop-go cycle that made long-range planning all the more difficult.

The travails of the British industry, especially those of British Leyland, were reflected in the cars that were produced. The Mini was followed by a number of models that used the same basic front-wheel-drive format. These were successful in engineering if not financial terms, but the last of the breed, the Austin Allegro, which was produced from 1973 to 1983 after a development process that had stretched out for five years, was plagued by dumpy styling, excessive interior noise, and dismal reliability. Even more tragic was the Rover SD1. Introduced in 1976 as a competitor for the likes of BMW and Mercedes, the Rover was attractively styled, but it soon gained a reputation for unreliability and poor build quality. Nonetheless, the name Rover still carried enough distinction that British Leyland was reorganized as the Rover Group in 1986. Two years later it was sold to British Aerospace at a price that was well below the value of its physical assets. The transfer of ownership did little to revitalize the firm that had once been the bedrock of the British automobile industry. By the early 1990s, the firm was producing only 400,000 cars annually and accounted for only 13 percent of the home market.

## AN OLIGOPOLY CRUMBLES

At the start of the last decade of the twentieth century, the survival of the American automobile industry was not in question, but Chrysler, Ford, and General Motors certainly were not in the position they enjoyed twenty-five years earlier. The United States had been a net exporter of automobiles in 1965, but now the situation was reversed as growing numbers of imported cars were shipped to the United States. The influx of imported automobiles was more than just a problem for the domestic producers, as it was the largest single source of a worsening trade imbalance for the United States as a whole.

For the domestic automobile manufacturers, the most important consequence of foreign-nameplate cars, both imported and domestically produced, was the loss of the cozy oligopoly that had prevailed for many years. Past competitors like Studebaker, Hudson, and Nash were gone, but in their place were much more formidable rivals like Toyota, Honda, and BMW. In some parts of the country, most notably California, Japanese and European imports outsold many long-established American makes by large margins.

Even highly successful firms faced many challenges in the twentieth century's last decade. Although there was no repetition of the energy crises of the 1970s, the long-term supply of motor fuel remained in doubt. Despite the great strides that had been made in automobile safety, in 1990 automobile accidents on average claimed a life every 11 minutes. Automotive emissions had been substantially reduced, but smog remained a persistent problem in many parts of the country. The spread of automobile-dependent suburbias in the United States and elsewhere continued to put a strain on land, resources, and the nerves of commuters. And through it all there was always the specter of new legislative initiatives aimed at the automobile and the problems that it engendered. Continued turbulence could be expected for everyone living in the world made by the automobile.

7

# At the Turn of a New Century, 1990–

◆

The concerns about safety, gasoline consumption, and air pollution that emerged with full force in the 1960s and 1970s seemed to portend a sharply diminished future for the automobile. Yet by the twentieth century's last decade, the automobile and the society it helped to create were showing impressive staying power. Technological advances had made cars safer, cleaner, and more fuel-efficient than could have been predicted thirty years earlier. Remarkably, all of these gains were accompanied by significant improvements in performance, although growing numbers of motorists seemed willing to sacrifice fuel economy, handling, braking, and overall safety by purchasing sport utility vehicles (SUVs), minivans, and pickup trucks. These vehicles helped American manufacturers to earn enormous profits, but competition from foreign manufacturers continued to eat into their share of the home market. In other lands, the growth of automobile ownership continued unabated, satisfying the desires of millions of new motorists, but also raising troubling questions about the effects of more cars on the global environment. Efforts to develop at least partial substitutes for private cars continued apace, but the dawn of a new century gave no indication of diminished interest in car ownership, much less that the automobile era was coming to an end.

## ADVANCES IN ENGINE TECHNOLOGY

By the early 1990s engineers had achieved remarkable successes in reducing automotive emissions while at the same time making the performance and drivability of cars even better than had been the case when pollutants were completely uncontrolled. The success of antipollution measures could be seen in the reduction of carbon monoxide and hydrocarbons emissions by 96 percent when compared to cars of the 1960s, and a reduction of oxides of nitrogen by 76 percent. In turn, cleaner cars made a key contribution to significantly better air quality. The improvement in air quality brought by cleaner cars was especially evident in Southern California, which for years had the dubious distinction of having the nation's most polluted skies. In 1977 the basin had suffered through 208 days in which federal health standards for one-hour ozone levels were exceeded. In 2001 there were only thirty-six such days (South Coast Air Quality Management District, 2004). Impressive progress in cleaning the skies could also be seen in the sharp drop in the number of first-stage smog episodes, from 121 in 1977 to none in 2001.

With emission-controlled cars spewing remarkably few pollutants into the air, much of the responsibility for poor air quality now rests with the relatively low number of "gross polluters" in operation. Gross polluters are defined as cars and trucks that generate at least twice the pollution-causing emissions as are allowed for that model of vehicle. According to estimates by the California Air Resources Board, although gross polluters constitute only 10 to 15 percent of the vehicles in California, they are responsible for more than 50 percent of the smog caused by cars and light trucks. Many gross polluters are old cars that are lacking in emissions-control equipment or are newer cars that have been poorly maintained. Older cars are not necessarily gross polluters, but they cannot be expected to be as clean as the current generation of emissions-controlled vehicles. At the same time, however, many old cars are special-interest vehicles that are driven infrequently and therefore are not a significant source of emissions.

As was noted in the previous chapter, early efforts to control vehicular emissions were unfortunately accompanied by serious performance losses and a host of "drivability" problems. What is impressive about the cars and small trucks currently being produced is that they combine remarkably clean exhausts with impressive overall performance. Even quite ordinary sedans are able to accelerate from 0 to 60 mph in 10 seconds or less, something that was beyond the capability of many sports and performance cars in the 1960s and 1970s. No less important, cars and trucks no longer suffer from the stalling, dieseling, and generally poor drivability that characterized the first generation of emissions-controlled vehicles.

These striking advances have not been the result of radically new engine technologies. Much of the improvement in engine performance can be attributed to the universal use of fuel injection, catalytic converters, electronic engine-management systems, and other technologies that can be traced back to the 1980s and earlier. The most innovative engine technology that took hold in the 1990s was the use of variable valve timing and lift, which was first used by Honda (under the name VTEC) for its high-performance Acura NSX two-seater. Engines equipped with this feature still use conventional poppet valves actuated by camshaft lobes, but the system overcomes an inherent limitation of conventional internal combustion engines. When it is turning at a high speed, an engine needs the intake and exhaust valves to remain open for a relatively long time to adequately fill the combustion chamber and completely scavenge the exhaust. But keeping the valves open for this length of time causes poor performance at lower engine speeds because some of the fuel-air mixture will be pushed through the open valves before it has been burned. In similar fashion, an engine turning at high speed benefits from a camshaft that lifts the valves a greater distance than is the case when the engine is rotating at lower speeds. Variable valve timing and lift get around these limitations by opening the valves with one camshaft lobe when the engine is running at low speeds, and a different lobe at higher speeds. This is done by hydraulically changing the rocker arms that connect the camshaft lobes with the valves so that one set of lobes actuates the valves at low engine speeds, and another set takes over at higher speeds. Changing the valve timing and lift thus allows an engine to idle smoothly and perform well at low speeds, and then to produce a surge of power at higher rotational speeds.

As an alternative way of combining good low- and high-speed performance, some engines do not use two sets of camshaft lobes for valve actuation; instead, the same camshaft lobes actuate the valves, but they do so earlier or later in accordance with the engine's needs. This is accomplished by a mechanism that advances the cam as the engine's rotational speed increases, and retards it as the engine slows. Because it does not change the time that the valves are open, it is not as effective as variable valve timing, but it requires fewer parts, which reduces manufacturing costs.

## BETTER HANDLING AND BRAKING

The cornering ability of cars, as measured by lateral acceleration, has also undergone a significant improvement. More sophisticated suspensions and stickier tires have given cars measurably better road-holding ability when compared with cars of the not-too-distant past. For example, while a high-performance Chevrolet Corvette two-seater from the late 1970s generated 0.726 g of lateral acceleration while cornering, by the year 2000 ordinary four-door sedans were capable of 0.8 g.

In the past, the exploitation of a vehicle's cornering ability depended on the skill of the driver. Safely taking a turn at high speed required deft control of the steering, throttle, and brakes while taking into account such factors as the vehicle's speed, its body roll, feedback from the steering, and the sensation of lateral acceleration. Today, however, stability-control systems can diminish the role of the driver as the car is able to control itself to a significant degree. With stability control, an electronic sensor monitors steering wheel movement in order to determine the direction the driver wants the vehicle to go, while other sensors monitor the rotational speed of the wheels, lateral acceleration, and the vehicle's yaw (the extent of movement on the vehicle's vertical axis). The information provided by these sensors is processed by an onboard computer to determine when the vehicle is on the verge of losing traction and going out of control. Should this be the case, the brakes are lightly applied to individual wheels, and in some systems engine torque is reduced. All of this occurs automatically and in a few thousandths of a second. For example, if the vehicle is in danger of oversteering (loss of traction of the rear wheels resulting in a tail-first slide), the rear wheel on the outside of a curve is braked. In the case of understeer (loss of traction of the front wheels), the braking force is applied to the rear wheel on the inside of a turn.

Stability control began to be available in the late 1990s and has met with a somewhat mixed reception among the purchasers of new cars. By 2003, 30 percent of the cars sold in Germany were equipped with this feature, but only 6 percent of the ones sold in the United States were so equipped. Stability control does seem to reduce accident rates; one study conducted by Toyota indicated that it reduced single-vehicle crashes by 35 percent and head-on crashes by 30 percent. Stability control is particularly valuable when applied to SUVs, which due to their high centers of gravity are more likely to go out of control and roll over than passenger cars.

The braking ability of cars has also improved through the use of disc brakes on all four wheels and tires that provide more grip. One innovative development in braking has been the use of antilockup technology, known as antilock braking system (ABS). It had long been known that a tire that continues to rotate while braking pressure is applied will stop more quickly and controllably than one that skids because the wheels have locked up. When confronted with a need to stop their vehicle as quickly as possible, many drivers have difficulty in applying maximum pressure on the brake pedal while avoiding wheel lockup. For this reason, drivers have been advised to pump the brake pedal, especially when stopping on slippery surfaces. ABS eliminates the need to skillfully modulate the application of the brakes by using sensors to detect when a wheel is about to lock up, momentarily releasing the brakes to restore traction, and then applying them once again, a process that may occur many times, with each application and release of the brakes lasting a fraction of a second.

There seems to be no question that ABS improves braking performance when compared with the braking done by average drivers. It would be thought, therefore, that cars equipped with ABS have a better safety record than those lacking this feature. At this point, however, accident statistics show no significant difference in accident rates of cars with and without ABS. Some of the reason for this lack of difference may be the tendency of drivers to continue with the practice of pumping the brakes, even though this diminishes ABS performance. Drivers may also be disconcerted by the normal pulsation of the brake pedal on ABS-equipped cars, which leads them to reduce the pressure they apply to the pedal. A third reason for the apparent inability of ABS to reduce accidents may be a phenomenon known as "risk homeostasis." This is the idea that safety features encourage people to take greater risks. In this case, better braking performance may cause people to drive faster, follow behind other cars at shorter distances, and in general drive in a less responsible manner. Whether or not this is actually true for the drivers of ABS-equipped vehicles is hard to determine, and more research will be needed before this assertion can be proved.

Features like ABS and stability control represent the continuation of a process that has characterized the evolution of automobiles for many decades. Technological advances like electric starters, synchromesh gearboxes, automatic transmissions, and now ABS and stability control have made driving easier and perhaps safer, but at the cost of greater complexity and expense. Moreover, these advances have less appeal for motorists who prefer a high level of interaction with the cars they drive than they have for people who look at their cars as a kind of appliance requiring minimal skill and involvement. And, in fact, average or below-average drivers can corner more rapidly in a vehicle equipped with stability control, but a skilled driver does better with the system switched off. As with many other advances in automotive technology, it can be argued that greater "user friendliness" has been accompanied by reduced driving skills and a more lackadaisical approach to operating a vehicle.

## AIRBAGS AND SAFETY

Despite all of the technological improvements that have contributed to making safer, better handling automobiles, accidents still happen, and efforts have to be made to diminish injuries and deaths when they occur. As was noted in the previous chapter, airbags became standard equipment during the 1990s, bringing with them the promise of reduced deaths and injuries. It became evident, however, that they were not the panaceas that some had hoped for. Although airbags may prevent or mitigate some crash-related injuries, they should not be viewed as substitutes for lap-and-shoulder belts, which continue to be the most effective means of preventing being ejected from a vehicle, and of avoiding the worst consequences of the "second collision" between a vehicle's occupants and its interior. The secondary role of airbags has been borne out by a Canadian study that concluded that seat belts were much more important in saving the lives of vehicle occupants seated in the front seats adjacent to the doors; from 1990 to 2000 it was estimated that seat belts saved 11,690 lives, while airbags saved 313 lives. Another study conducted by the National Highway Safety Administration in the United States came to somewhat more positive conclusions about the value of airbags, finding that by themselves they reduced the fatality risks by 14 percent, while the use of lap-and-shoulder belts resulted in a 45 percent reduction. Used together, both restraint systems lowered the risk of a fatality by 50 percent.

Although it has become evident that airbags do not provide adequate protection by themselves, safety officials in the federal government initially expected that most drivers could not be convinced to use their belts; consequently, airbags were designed to protect unbelted adult drivers. This resulted in the fitting of airbags that inflated with explosive force at speeds of up to 200 mph (320 kph). Airbags of this sort posed a lethal threat to children and short drivers who had to sit close to the wheel. Within a few years of their widespread installation, it became apparent that airbags were literally killing people; the National Highway Traffic Safety Administration estimated that by mid-2003, 231 people (144 of them infants and children) in the United States had lost their lives as a result of airbag deployments.

The death toll caused by airbags has led to a rethinking of the role of passive restraints in automobile safety. It is now assumed that the majority of drivers use their seat belts (in fact, about 70 percent do so); consequently, airbags do not have to bear the entire burden of restraining a vehicle's occupants in the event of a crash. This has allowed a reduction in the force with which airbags deploy, making them less dangerous for children and drivers in close proximity to the steering wheel. Even so, they still pose a risk. Drivers are now advised to grip the lower portion of the steering wheel in order to lessen the risk of broken arms in the event of an airbag inflation. Most importantly, rear-facing infant seats should never be put on the front seat, and children under the age of 12 should always sit in the rear seat, away from a possible airbag deployment.

The continued danger posed by airbags has motivated efforts to come up with safer designs. "Advanced frontal airbags" automatically determine their inflation force by taking into account a number of parameters such as the size of the seat's occupant, whether a seatbelt is being used, the position of the seat, and the severity of the anticipated crash. This new generation of airbags is being phased in for cars and trucks sold in the United States, and they will be mandatory for all vehicles produced after September 1, 2006.

## TRUCKS FOR NONTRUCKERS

For many decades, the production and use of private vehicles centered on automobiles. Trucks for the hauling of goods emerged during the first decade of the twentieth century and grew in substantial numbers in the years that followed, but most of them were used for commercial purposes, not for everyday driving. Some trucks were built on automotive chassis, and some cars like the Model T Ford were equipped with pickup beds, but for the most part trucks and cars constituted separate species that were intended for different purposes.

The distinct roles of cars and trucks began to blur as pickups, vans, and sport utility vehicles increasingly took the place of conventional passenger cars. After decades of comprising only 10 percent of vehicle sales in the United States, light trucks accounted for 20 percent of vehicle sales in 1974, and then increased to 30 percent in 1987 and to 40 percent in 1994 (Rubenstein 2001, 235–36). By the year 2001, the score was about even when 8,655,073 cars and 8,522,372 trucks were sold in the United States (*Market Data Book*, 2002). By this point, the full-sized pickup trucks made by Chevrolet and Ford were firmly ensconced as the largest selling vehicles in the United States, and the Ford Explorer SUV and the Chrysler minivan perennially registered among the top ten largest selling vehicles.

The popularity of light trucks has had some negative consequences. On average, these vehicles delivery worse fuel economy than cars. Increased numbers of fuel-inefficient light trucks have been an important source of rising gasoline consumption, as Table 7.1 indicates.

One of the most striking automotive trends in recent years has been the popularity of SUVs in the United States. During the 1990s sales of large SUVs like the Ford Excursion and the Chevrolet Suburban went from

**Table 7.1**
**Car and Truck Fuel Consumption**
**in Billions of Gallons**

| Year | Cars | Vans, Pickups, and SUVs |
|---|---|---|
| 1970 | 67.8 | 12.3 |
| 1980 | 70.2 | 23.8 |
| 1990 | 69.8 | 35.6 |
| 1999 | 73.2 | 52.8 |

U.S. Census Bureau, *Statistical Abstract of the United States,* 2001, Table 1105.

60,000 to 600,000, while 1.6 million medium-sized SUVs like the Jeep Grand Cherokee were sold annually during the last years of the decade (Rubenstein 2001, 241).

SUVs appealed to many drivers because they had the interior room of minivans and station wagons while not suffering from the dorky image presented by these vehicles. Although the vast majority of SUV drivers confined their off-road excursions to nothing more adventurous than a dirt parking lot, their rugged, outdoor image has been a big part of their appeal. No less important, people bought SUVs because they were heavily promoted by the automobile companies that made them. The manufacturers had good reason to aggressively market SUVs. For one thing, SUVs were subject to looser fuel economy measures; whereas manufacturers had to attain a corporate average fleet economy (CAFE) standard of 27.5 mpg for the cars they produced, their SUVs had to average only 20.7 mpg. SUVs have also brought in massive profits for their manufacturers. Many of them are built on pickup truck chassis, and therefore require less new tooling than totally new vehicles. At the same time, the high demand for SUVs allows their makers to keep their prices and consequently their profits at a high level. Some large SUVs like the Cadillac Escalade reportedly brought in profits of $15,000 per unit. By contrast, an automobile that costs about as much to manufacture may bring in only a few hundred dollars of profit.

Perhaps the most problematic aspect of the SUV boom centers on safety, or the lack of it. According to critics, the high center of gravity typical of SUVs makes them less stable than passenger cars. Sudden maneuvers, such as swerving to avoid a dog that has darted into the road, may result in the vehicle flipping over. Statistics from the late 1990s seem to bear out the presence of this inherent danger. Each year there were ninety-eight rollover accidents for every million miles traveled by SUVs; in contrast, the rollover rate for all types of vehicles was only forty-seven per million miles.

This should be a particularly troubling statistic for SUV owners because rollover accidents have a higher rate of injuries and fatalities than other kinds of accidents. Although in recent years, only 3 percent of the motor vehicle accidents in the United States were rollovers, they accounted for one-third of all fatalities for vehicle occupants. For SUV drivers and their passengers, the greater propensity to be involved in a fatal rollover accident more than offsets the safety benefits conferred by being in a large vehicle. As the head of the National Highway Traffic Safety Administration noted in 2003, when compared to an occupant of a regular passenger car, an SUV occupant was three times more likely to be killed as a result of a rollover accident (*Wall Street Journal* January 15, 2003). Moreover, the safety problems posed by SUVs involve more than their occupants. When fatalities to the drivers and passengers of other vehicles involved in collisions with SUVs are taken into account, SUVs are on average 30 percent more dangerous than large cars and 25 percent more dangerous than midsized ones (Ross and Wenzel 2002, 5).

Pickup trucks are even more lethal than SUVs in the event of an accident involving another vehicle; pickup-to-car collisions result in twice as many deaths as car-to-car crashes (Ross and Wenzel 2002, 6). One particular kind of collision between pickups and cars is especially deadly for the occupants of the latter. In the kind of collision where a full-size pickup truck hits a passenger car broadside, the car's occupants are twenty-six times more like to die as the pickup's occupants, a ratio three times greater than car-to-car crashes of this type (*Wall Street Journal* January 15, 2003).

In evaluating the relative safety of different kinds of vehicles, it is of course necessary to take into account a number of factors other than the size and design of a car or truck. Of these, the characteristics of the typical driver of a vehicle is one of the most significant influences on its safety record. For example, one of the key reasons that sports cars have a considerably higher rate of fatal accidents than luxury cars is that they are driven by younger and presumably more aggressive drivers. It is also likely that the relatively high incidence of fatalities among the occupants of pickup trucks is in part due to their being driven on rural roads where conditions may be more hazardous than urban and suburban roads.

Disentangling all of the factors that affect the relative safety of different types of vehicles is a difficult task, but one systematic study indicated that some broad generalizations can be made: on average, SUVs are no safer than large and medium-sized cars; sports cars and pickup trucks are the most dangerous vehicles on the road, and minivans are the least dangerous; and domestic American vehicles are less safe than vehicles produced by Japanese and European manufacturers (Ross and Wenzel 2002). At the same time, it is important to recognize that the behavior of drivers remains the most important determinant of highway safety. The great majority of accidents can be attributed to driver error of some sort, and driving while under the influence of alcohol is the greatest single cause of fatal accidents. About 30 percent of vehicle fatalities involve a driver with a blood alcohol level of 0.10 grams per deciliter, the standard for impairment used by most states (*Statistical Abstract of the United States* 2001).

## A GLOBALIZED INDUSTRY

The increasing sophistication of cars and trucks has greatly increased the costs of bringing them to market. Designing, developing, testing, and preparing to manufacture an all-new design can easily cost several billion dollars. One of the surest ways to bring down the production costs of individual vehicles is to manufacture large numbers of them in order to benefit from economies of scale, as was dramatically illustrated by the Ford Motor Co. many decades ago. Today, the pursuit of very large production runs has been one of the motivations for the continuation of another longstanding practice: the merger of smaller firms into larger ones.

What has made recent mergers particularly striking has been their global scale. This of course is not a new phenomenon; as we have seen, Ford and General Motors have had overseas operations and affiliates for many decades. In the 1990s, however, the shoe was on the other foot when Germany's Daimler-Benz acquired one of the American Big Three, Chrysler, to create DaimlerChrysler. Consummated in 1998, this was the biggest merger of industrial firms ever, one that united one giant firm with market capitalization of nearly $28 billion and 121,000 employees (Chrysler) and another with an even larger market capitalization of more than $50 billion and 300,000 employees (Daimler-Benz).

Massive though it was, the DaimlerChrysler tie-up was part of an even larger international collaboration. Prior to being taken over by Daimler-Benz, Chrysler had extended its international reach through its ownership of 13 percent of Mitsubishi Motors. In 1988 a joint venture of the two firms began to produce small cars in the United States. Chrysler's partial ownership of Mitsubishi went back to 1970, the same year that Ford had acquired 20 percent of Toyo Kogyo, the manufacturer of Wankel-powered Mazda cars. By the 1990s Ford had acquired a 37.5 percent share of Mazda (as Toyo Kogyo had been renamed), giving it a controlling interest in the Japanese firm.

By the beginning of the twenty-first century, purely domestic automobile industries had virtually disappeared amidst a wave of cross-national acquisitions, mergers, and joint ventures. Along with its controlling interest in Mazda, Ford directly owned a number of renowned European car makers: Jaguar, Volvo, Aston Martin, and Land Rover. General Motors, which long had a presence in Europe through its ownership of Vauxhall in Britain and Opel in Germany, extended its global reach through its ownership of sizeable shares of Japan's Isuzu and Subaru, Italy's Fiat, and Korea's Daewoo, as well as complete ownership of Sweden's SAAB. Global reach in the automotive industry also emanated from France, as Renault gained a controlling interest in Nissan and saved it from imminent bankruptcy.

Volkswagen, which had attempted in the late 1970s to follow the Japanese example of manufacturing cars in the United States, abandoned that venture in 1988 and redirected its multinational reach to Europe. In 1986 it acquired the Spanish firm SEAT, to which was added the Czech firm Skoda in 1990. It also maintained a presence in North America through its production facility in Mexico, as did four other multinational producers (Chrysler, Ford, General Motors, and Nissan). Another cross-border acquisition involved the sale of two quintessentially British firms, Rolls-Royce and Bentley, to BMW and Volkswagen, respectively. At the other end of the scale, BMW acquired Britain's Rover Group (Austin Rover prior to 1986) in 1988 in the hope of complementing its high-priced sports sedans and luxury models with a lower priced line of front-wheel-drive cars. Unable to make a profit, BMW virtually gave the operation away to a British investment group in 2000.

When they succeeded, these international linkages served as the basis for a substantial amount of parts sharing and the consequent realization of economies of scale. Major components that were relatively easy to substitute, like engines and transmissions, were installed in different makes of cars, such as the use of Mitsubishi V-6 engines in Chrysler minivans. In order to further benefit from economies of scale, manufacturers have gone beyond component sharing and increasingly use common platforms as the basis for a variety of cars and light trucks that could be manufactured all over the world. To take one example, within the Ford empire, a platform developed by Mazda is the basis for that firm's small car, another small car bearing the Volvo name, and a compact SUV sold as a Ford.

## THE MOTORIZATION OF THE THIRD WORLD

The great majority of the world's automobiles ply the roads of North America, Europe, and Japan. In recent years a few other countries, notably South Korea, have experienced a rapid growth of automobile ownership. In the case of Korea, the expansion of the car population was almost entirely due to the expansion of the domestic industry, which for a number of years was completely insulated from foreign competition by government policies. As it turned out, growth was too rapid, as an excessive number of firms tried to get into the business. A shakeout ensued with Hyundai emerging as the dominant producer, but, significantly, it could not maintain its status as a completely independent manufacturer; 15 percent of its stock is owned by DaimlerChrysler, and another 5 percent is owned by Mitsubishi. Daewoo, the car-making division of the third-largest Korean corporation, could not keep up the competitive pace; its bankruptcy led to a two-thirds ownership stake by General Motors. At the same time, however, all of the vicissitudes of the Korean automobile industry should not obscure the fact that Korea has emerged as the world's fifth largest automobile manufacturer with a production of more than 2.7 million cars in 2003, and that it is well on the way toward being a country with universal automobile ownership.

India, the world's second largest country in terms of population, also tried to maintain an independent automobile industry, but its chief products were copies of 1950s-era Austins and Fiats that sold in relatively small numbers. In the 1980s India underwent a significant shift in economic policy when its government began to allow joint ventures between domestic and foreign firms. In the automotive industry, the most important of these was an alliance between Maruti-Udyog and Suzuki. The joint venture manufactured a minicar with an 800-cc engine that sold for the equivalent of $6,000. The car, known as the Maruti, captured 80 percent of the domestic market during the 1990s, greatly aided by the very high tariffs imposed on imported automobiles.

The greatest prospect for increased automobile production can be found in China. With its 1.3 billion people, even at the present ratio of one privately owned car for every 130 people, it constitutes a significant market, one that the world's automobile manufacturers are scrambling to enter. More than 2 million cars were sold in China in 2003, making it the world's fourth largest market, behind only the United States, Japan, and Germany (National Bureau of Statistics of China 2002). This trend will most likely continue. Benefiting from one of the world's fastest growing economies, a significant portion of China's population is reaching income levels that make the purchase of a car possible, and there seems no reason to doubt that the number of cars on China's roads will increase dramatically, adding substantially to the world's car population. Reaching the 6 percent car-to-family ratio achieved by Japan in the mid-1960s will add 30 million cars, more than existed in the United States in 1946. And with a car-to-person ratio approaching today's Japan or Europe, China would have the world's largest automobile population by far.

The Chinese government has been determined to ensure that most of the expanding domestic market will be supplied by domestic producers. In 1996 it designated the Chinese car and truck industry a "pillar industry" that qualified for preferential treatment. As with other countries in an early stage of developing an automobile industry, the government of China encouraged the formation of joint ventures with foreign firms, but it did everything in its power to prevent these companies from dominating the industry. At this point, these firms are content to share the profits accruing from the expanding Chinese market for private automobiles. It is likely, however, that at some time in the future the Chinese automobile industry will not be satisfied with meeting domestic needs and will become a major competitor in the global automobile market.

At the beginning of the twenty-first century, the world's population of cars and light trucks totaled about 550 million, while annual output was in the neighborhood of 50 million. All that it would take to double world car production is to have a mere 5 percent of the households of India, China, Indonesia, and other Asian countries become car owners (Rubenstein 2001, 355). The expansion of the automobile industries of the industrially underdeveloped countries is good news for populations who still rely on bicycles, pedicabs, and oxcarts for their daily transportation, but it is also a cause for concern. A major expansion in automobile production and ownership will put a substantial strain on the environment of these nations and the world as a whole. If this should come to pass—and there is no reason to think that it won't—the world's petroleum supplies, although in no danger of immediate depletion, would be drawn down more rapidly, and the tightening supplies would inevitably increase the price of gasoline worldwide. A large vehicle population, even when equipped with the most up-to-date emissions controls, would be a major source of air pollution and would contribute significantly to carbon dioxide emissions, widely considered to be a prime cause of global warming. It is understandable that people in the developing world aspire to a level of consumption that was reached in the industrially developed countries decades ago, but it will not be a painless achievement, and some of that pain will be felt beyond the borders of these countries.

## REINVENTING THE AUTOMOBILE

A large portion of the present and future problems caused by widespread automobile ownership can be summed up in two words: "air pollution." As we have seen, a number of technological fixes have made internal combustion engines much cleaner, but it is still hoped that cars and light trucks will eventually have power sources that produce no emissions whatsoever. Given the state of technology in the present and recent past, there is only one reasonably practical zero-emission vehicle, the battery-powered electric car or truck.

Although they had a substantial clientele during the automobile's infancy at the beginning of the twentieth century, electric cars rapidly faded from the scene. In an effort to revitalize the sale of electric cars, in the early 1990s the California Air Resources Board mandated an ambitious timetable for the production of zero-emission vehicles: 2 percent of a manufacturer's sales in the state by 1998, 5 percent in 2000, and 10 percent in 2003. These standards were subsequently adopted by Massachusetts, New York, and the District of Columbia. The zero-emissions mandate was subsequently relaxed, but manufacturers still had to prepare for a time when they would have to manufacture and market significant numbers of electric vehicles. The most publicized effort was mounted by General Motors. Their EV1 was a rakish two-seater with impressive acceleration, but even with advanced electronics technology, the EV1 still suffered from the same drawback that plagued electric cars a century earlier: lack of range. General Motors claimed that the EV1 could travel up to 100 miles between battery recharges, but stop-and-go traffic coupled with the use of lights, radio, and air conditioning dropped its range to 60 miles or less. Even so, the EV1 aroused considerable enthusiasm among those who leased it (the car was not available for sale), and many expressed both sadness and outrage when General Motors refused to renew their leases and the fleet of a few hundred cars was taken off the road in 2003.

By this time the Air Resources Board had revised their zero-emissions rules. One provision of the revised mandate gave manufacturers credit for producing hybrid cars, vehicles that combine an electric motor with a conventional internal combustion engine. As a result, it is likely that more than 400,000 hybrids will be sold in the state from 2005 to 2010, double the projected number envisaged before the new rules were enacted. While battery-powered electric cars failed to gain a commercial toehold, hybrids have shown considerable technical and commercial promise, and they are likely to be produced in large numbers in the years to come.

As with many other seemingly radical innovations, powering a car through the combination of an electric motor with a gasoline engine has a lineage that goes back many decades. In 1900 Ferdinand Porsche (the same Dr. Porsche who designed the Volkswagen Beetle and whose name is commemorated in the legendary family of sports cars) created an automobile that was propelled by four electric motors mounted on the car's wheel hubs. The motors received their current from an electrical generator that in turn was powered by internal combustion engine. In the years that followed, several other firms built cars that combined electrical with gasoline power, but never in appreciable numbers.

Gasoline-electric hybrids made a comeback in the 1990s when a few manufacturers began to use the technology as a way of producing more fuel-efficient automobiles. The first of these was the Honda Insight. The Insight was a parallel hybrid, which means that it was propelled by a 10 kW electric motor, a 67-hp, three-cylinder gasoline engine, or both working together. The gasoline engine was the primary source of power, but the electric motor kicked in when more power was required, such as accelerating from a stop. Although the electric motor used with the Insight had limited power, like all electric motors it put out maximum torque from the moment it started rotating, so it was particularly useful when acceleration was required. As with all hybrids, the Insight employed regenerative braking. In conventional cars, slowing down by applying the brakes results in nothing but the generation of heat, whereas the application of the brakes in a hybrid car causes the electric motor to act as a generator, so some of the kinetic energy that otherwise would be lost can be used to recharge the car's battery.

The Toyota Prius, the second hybrid sold to the general public, had a 70-hp internal combustion engine and a 44-hp electric motor that developed a whopping 258 ft-lb of torque at up to 400 rpm. The Prius used a planetary gearbox to allowed three different modes of running: gasoline engine only, electric motor only, or both working together. When accelerating from a stop, the Prius used only the electric motor until a speed of about 15 mph (24 kph) was reached, at which point the gasoline engine started up. It then ran at a constant speed; acceleration was accomplished by changing the speed of the electric motor as well as the generator that supplied current to it.

These first-generation hybrids had excellent fuel economy (40–45 mpg) because they employed relatively small internal combustion engines. Speed and acceleration would suffer if these engines had been the sole source of power, but with the assistance of an electric motor the cars were able to perform about as well as similarly sized cars with larger, less fuel-efficient engines. Even so, they did not represent a massive improvement over conventional cars of similar size. Greater gains in overall fuel savings will be achieved when hybrid power systems are used for large vehicles. GM's hybrid pickup trucks are expected to get at least 15 percent better fuel economy than conventional models, and since these vehicles use gasoline at the rate of about 14 miles per gallon, even a relatively slight improvement would significantly reduce the nation's gasoline consumption if large numbers were on the road.

Hybrid cars offer better fuel economy and lower emissions, albeit with more complication, than cars with conventional sources of power. A more radical approach to automotive power is the fuel cell. The basic concept of the fuel cell goes back to its invention by William Grove in 1839. Its operation can best be described as reverse electrolysis; hydrogen or a fuel containing hydrogen goes through a process that causes it to lose electrons, resulting in a flow of electric current. Fuel cells have two important advantages over internal combustion engines: they operate more efficiently and they produce fewer emissions. A fuel cell running on hydrogen operates three times more efficiently than an internal combustion engine running on hydrogen, and a fuel cell's "exhaust" consists of water along with small quantities of carbon dioxide and carbon monoxide.

Despite these advantages, the eventual adoption of the fuel cell as an automotive power plant is hardly a certainty. At this point, there are four major obstacles to a widespread conversion to fuel cell–powered vehicles. First is the large amount of space they occupy, leaving little interior room for people and cargo. Second is the cost of manufacturing them along with the auxiliary equipment they require. Prototype fuel cell cars have been made by a number of manufacturers, and in 2003 a small fleet produced by Honda was being operated by the City of Los Angeles. With a production cost of $2 million per car, far more than the city paid for their use, the program was meant to provide real-world experience with fuel cells, and nobody expects that members of the driving public would pay anything like that sum. As has happened many times in the history of the automobile, series production would likely bring costs down, but it is not certain that they can be lowered to the point where fuel cells would be competitive with conventional engines. The third problem is the present lack of a supportive infrastructure. Cars powered by internal combustion engines are the recipients of over a century of infrastructure development, fueling stations and repair shops in particular. Developing parallel facilities, especially where the storage and distribution of hydrogen is concerned, would take time and billions of dollars of investment funds. Finally, there is the matter of hydrogen production. Although hydrogen is the most abundant element in the universe, it is not easily obtained on planet Earth. Extracting it from natural gas is the most common and least expensive way of obtaining hydrogen. A fuel cell car running on hydrogen obtained from natural gas would use about as much energy as a car with an internal combustion engine using gas as fuel. It would, however, be more expensive, and the situation will likely worsen in the future, given the finite supply and rising cost of natural gas.

Hydrogen also can be obtained through the electrolysis of water. The problem here is not the availability of water, but the cost, in terms of both money and energy, of separating hydrogen and oxygen molecules. Electricity derived from solar power is attractive from an environmental standpoint, but it costs ten times more than electrical generation based on coal or natural gas. The price of solar-generated electricity may decline, but massive amounts of land and capital investment would still be required if electrolysis through the use of solar energy were to be employed on a large scale. Nuclear power could also be used for the electrolysis of water, but this might entail trading one environmental problem for another.

## A POSTAUTOMOBILE WORLD?

For some critics of the automobile, no amount of tinkering with its technology will be sufficient. Even if cars produced no pollution and caused no deaths or injuries, they would still be a problem. According to their detractors, automobiles devour great amounts of land for roads and parking, absorb natural resources in vast quantities, and promote an excessively privatized existence. What is needed, therefore, is a system of public transportation that sharply limits or even eliminates the need for private automobiles.

The historical decline of public transportation in the United States remains a contentious issue. The often-repeated tale that General Motors, Firestone Tire and Rubber, and Standard Oil conspired to destroy public transportation in numerous American cities is little more than an urban legend, but it cannot be denied that beginning in the 1920s preference for travel by car was responsible for the steady decline of passenger rail service, interurban electric railroads, and urban bus lines. It also has to be recognized that governmental policies that overregulated the railroads, sponsored the interstate highway system, and encouraged suburban growth also contributed mightily to the erosion of public transportation. In contrast to the situation in many parts of the United States, thriving public transit systems operate in Europe and Asia, but their success has entailed large public subsidies.

Efforts by the federal government to redress the imbalance between public and private transportation in the United States began in 1973, when the U.S. Congress allowed individual states to use a portion of their Interstate Highway Trust fund allotments for capital expenditures on public transit. In 1978 Congress increased federal support by raising to 85 percent the federal share of funding for public transit projects, considerably more than the two-thirds share originally enacted. In 1982 Congress raised the gasoline tax by 5 cents, of which 1 cent was dedicated to transit projects. In the years that followed, 20 percent of all increases in the federal gasoline tax was allocated for public transit.

By the early 1990s, governments at all levels were spending lavishly on public transit. In 1992, for example, the federal government contributed $3.7 billion, while the contributions of state and local governments were more than five times greater at $18.7 billion. All in all, by this point, federal, state, and local governments were allocating to public transit nearly 25 percent of government expenditures on highways and transit (Dunn 1998, 108). Some of these funds went to projects that absorbed enormous sums of money, most notably the initial four-mile stretch of the Los Angeles subway, which cost $300 million per mile and subsidized each passenger trip to the tune of $21.

Governmental financial help was no less crucial to the survival of intercity rail in the United States. Carrying passengers by rail probably never was a very profitable activity, and during the post–World War II era it had become a burden that the financially troubled railroads could no longer endure. In 1970 they were relieved of that money-losing enterprise when Congress created the National Railroad Passenger Corporation, commonly known as Amtrak. In 1995 the federal government provided $393 million to help cover operating costs and $230 million in capital investment funds. Despite a large infusion of federal money over the years, Amtrak's financial situation remains shaky, its ridership has not increased appreciably, and above all, it has done little to diminish long-distance automobile travel; cars account for 80 percent of trips of 100 miles or more, with air travel responsible for most of the remainder.

Although railroads, subways, light rail systems, buses, and other mass transit modes are properly considered as key parts of a balanced transportation system, a hard fact remains: mass transit requires concentrated masses of passengers, but in the United States and in many other parts of the industrially developed world the dominant trend has been toward reduced population density due to the dispersal of residences, places of business, and shopping and entertainment facilities. One can argue over the exact place of the automobile in motivating low-density living, but it certainly has not been the only factor. The most that can be said with complete conviction is that widespread automobile ownership has been both cause and effect of the decentralized, suburbanized, and home-centered mode of life prevalent today.

In 1995 a symbolic threshold was crossed when the population of cars and light trucks in the United States equaled the number of licensed drivers. Another milestone was reached in the year 2003 when the number of cars in the average American family unit exceeded the number of licensed drivers in the house. At this point, the triumph of the automobile seemed to be complete. But at the same time there is no denying that the problems engendered by the automobile have grown apace. At the very least, it can be said that the automobile has been too successful for its own good. Traffic congestion is a major problem in most urbanized areas, and it usually ranks high on the list of citizens' complaints about the quality of daily life. So problematic has congestion become that some municipalities, most notably London, have taken the extreme step of charging substantial fees to vehicles entering the central city. Another major irritant, both literally and figuratively, is air pollution. Although impressive progress has been made in the reduction of automotive emissions, the relentless expansion of the world's vehicle population threatens to nullify these gains and even reverse them. And even if technological advances produced cars with no smog-forming emissions whatsoever, there remains the inescapable fact that the combustion of any carbon-based fuel results in the production of carbon dioxide, which in all likelihood is a major contributor to global warming and its attendant potential for serious environmental disruption.

All of these automobile-based problems may be rendered moot if future automobile use ends up being sharply constricted by the depletion of petroleum supplies and a consequent scarcity of gasoline and diesel fuel. Assessments of current oil resources and their future trajectory are highly imperfect; some experts predict significant declines by the end of the first decade of the present century, while others see this happening much later. In any event, it does seem fair to state that economically accessible oil supplies will undergo a significant decline at some point in the lives of many readers of this book. It is possible that the use of substitute fuels like hydrogen or some variety of alcohol will render the exhaustion of oil supplies irrelevant, but technical feasibility is not the same thing as commercial practicality, and at this point it is unlikely that any alternative fuel will equal or surpass petroleum-based fuels in energy density, transportatibility, and low cost. Perhaps higher fuel costs will not severely restrict automobile ownership and use, but they will require sacrifices in other areas of consumption if the present extent of automobile use is to be maintained.

While the costs of car ownership and operation are destined to increase for individuals and for society as a whole, in all likelihood they will continue to be borne with little complaint because few devices have offered as large and varied a package of individual benefits as has the automobile. Convenient and reliable transportation remains the chief justification for having a car, but the automobile's unique ability to confer privacy, power, status, and fascination are no less important. At the most fundamental level, the automobile has given its owners a freedom that is not available to nondriving populations. Ironically, however, the very freedom promised by the automobile has been at least partially offset by a parallel set of restrictions— traffic laws and regulations; congestion; massive financial demands in the form of depreciation, maintenance, repairs, and insurance; and the taxes required to maintain and expand the vast infrastructure of roads, highways, and parking lots that sustains the automobile. To these must be added the indirect but all-too-real costs of accidents, air pollution, noise, and visual blight that are caused by widespread automobile ownership. The automobile has driven a hard bargain, but so far it has been one that we have accepted willingly, even enthusiastically.

# Glossary

**Catalytic converter.** A large canister placed in a car's exhaust system ahead of the muffler. In it, a catalyzed chemical reaction converts carbon monoxide into carbon dioxide, oxides of nitrogen into free nitrogen and oxygen, and unburned hydrocarbons into water vapor and carbon dioxide.

**Compression ratio.** The ratio of the volume of the interior of an engine's cylinder when the piston is at the bottom of its stroke, plus the combustion chamber, to the volume of the combustion chamber alone. In general, a high compression ratio makes an engine run more efficiently, but knocking and pinging will result if it is excessively high.

**Diesel engine.** A type of internal combustion engine that uses the compression of the air in the combustion camber to generate the heat that ignites a mixture of fuel and air.

**Disc brakes.** A means of stopping a vehicle in which one or more sets of pads squeeze a disc that is bolted to a car's wheel. Disc brakes provide more even braking power and are less likely to fade than drum brakes.

**Four-stroke engine.** A type of internal combustion engine that uses four strokes of the piston (two up and two down) to complete one cycle of operation. The four strokes are intake, compression, ignition and power, and exhaust.

**Fuel injection.** A means of getting fuel into an engine's combustion chamber.

Powered by a high-pressure pump and dispensing fuel through a nozzle with a very small aperture, a fuel injection system provides more precise delivery of fuel than a carburetor.

**Hydraulic brakes.** A means of applying brakes in which a fluid instead of a mechanical connection is used to apply pressure to the brake shoes or pads.

**Independent front suspension.** Unlike a solid-axle front suspension, in which the two front wheels are connected by a beam, with this type of suspension both wheels are free to move independently of each other.

**Internal combustion engine.** A type of engine in which heat and pressure are created within the engine itself, rather than being delivered from an external boiler, as is the case with a steam engine.

**Leaf springs.** A suspension medium that uses one or more long, blade-shaped leaves for each wheel. The leaves may be deployed in the shape of an ellipse or a semi-ellipse.

**Magneto.** An electrical device that supplies a timed electric current to the spark plugs. Current is produced when the movement of a magnetized rotor produces a changing electrical field in the wire windings of a fixed stator. Alternatively, the stator may be the source of the magnetic field.

**Overhead valve engine.** A type of engine in which the intake and exhaust valves are situated in the cylinder head and open into the top of the combustion chamber. The valves are opened by a camshaft that lies over them, or by an arrangement of rocker arms and pushrods actuated by a camshaft below. This configuration allows better combustion chamber design than a sidevalve arrangement, but requires more moving parts.

**Planetary transmission.** A type of transmission in which one gearwheel (the "planet" gear) "orbits" around another rotating gear (the "sun" gearwheel), while at the same time engaging the inner teeth of a surrounding ring gear. Locking in different gears allows changes in gear ratios and consequent changes in the speed of a car's wheels relative to the speed of its engine. This arrangement was used for the transmission of the Ford Model T and continues to be used for automatic transmissions today.

**Power assistance.** The use of power supplied by a car's engine to reduce the physical effort needed to operate steering or brakes.

**Sidevalve engine.** A type of engine in which the intake and exhaust valves are directly actuated by a camshaft situated in the lower part of the engine alongside the crankshaft. In this arrangement, the intake of a fresh charge and the exhaust of a spent charge take place on one side of the combustion chamber. Engines of this type are also known as "flatheads."

**Supercharging.** A means of increasing an engine's power by forcing air and fuel into the combustion chambers. This can be done by an engine-driven impeller, or by using the engine's exhaust to drive a small turbine (turbocharging).

**Synchromesh transmission.** A type of transmission that uses synchronizers to match the rotational speeds of the gears being engaged, thereby making it easier to shift from one gear to another.

**Torque.** The measurement of an engine's twisting force, or more generally what is obtained when a force is exerted over a distance. Torque is conventionally measured in newton-meters or pound-feet. The horsepower of an engine is obtained by multiplying torque in pound-feet by the engine's rotational speed, and dividing the product by 5252.

**Wankel rotary engine.** A type of engine that dispenses with conventional reciprocating pistons by using a rotor shaped like a triangle with bulging sides that moves eccentrically within a housing to perform the four basic functions of an internal combustion engine: intake, compression, ignition and power, and exhaust.

# Bibliography

American Automobile Manufacturers Association. *Automobiles of America*. Detroit: Wayne State University Press, 1962.

Anderson, Robert. *Fundamentals of the Petroleum Industry*. Norman: University of Oklahoma Press, 1984.

Association of Licensed Automobile Manufacturers. *Hand Book of Gasoline Automobiles, 1904–1906*. New York: Dover, 1969. Reprint of 1904–1906 editions.

Bardou, Jean-Pierre, Jean-Jacques Chanaron, Patrick Friedenson, and James M. Laux. *The Automobile Revolution: The Impact of an Industry*. Chapel Hill: University of North Carolina Press, 1982.

Berger, Michael. *The Devil's Wagon in God's Country: The Automobile and Social Change in Rural America, 1893–1929*. Hamden, CT: Archon Books, 1979.

Bird, Anthony. *Antique Automobiles*. New York: E. P. Dutton, 1967.

Brandon, Ruth. *Auto Mobile: How the Car Changed Life*. London: Macmillan, 2002.

Brilliant, Ashleigh. *The Great Car Craze: How Southern California Collided with the Automobile in the 1920s*. Santa Barbara, CA: Woodbridge Press, 1989.

Carrieri, Raffaele. *Futurism*. Leslie van Rennselaer White, trans. Milano: Edizioni de Milione, 1963.

Chandler, Alfred D., Jr., ed. *Giant Enterprise: Ford, General Motors, and the Automobile Industry*. New York: Harcourt, Brace & World, 1964.

Church, Roy. *The Rise and Decline of the British Motor Industry*. Cambridge: Cambridge University Press, 1995.

Clutton, Cecil, and John Stanford. *The Vintage Motor Car*. London: B. T. Batsford, 1961.

Cray, Ed. *Chrome Colossus: General Motors and Its Times*. New York: McGraw-Hill, 1980.

Dunn, James A., Jr. *Driving Forces: The Automobile, Its Enemies, and the Politics of Mobility*. Washington, DC: Brookings Institution Press, 1998.

Editors of *Consumers Guide*. *Cars of the 30s*. New York: Beekman House, 1978.

———. *Cars of the 40s*. New York: Beekman House, 1979.

———. *Cars of the 50s*. New York: Beekman House, 1980.

Flink, James J. *America Adopts the Automobile, 1895–1910*. Cambridge, MA, and London: MIT Press, 1970.

———. *The Automobile Age*. Cambridge, MA, and London: MIT Press, 1988.

———. *The Car Culture*. Cambridge, MA, and London: MIT Press, 1975.

Ford, Henry, with Samuel Crowther. *My Life and Work*. Garden City, NY: Doubleday, 1922.

Greenleaf, William C. *Monopoly on Wheels: Henry Ford and the Selden Automobile Patent*. Detroit: Wayne State University Press, 1961.

Hounshell, David. *From the American System to Mass Production, 1800–1932: The Development of Manufacturing Technology in the United States*. Baltimore and London: Johns Hopkins University Press, 1984.

Hyde, Charles K. *Riding the Roller Coaster: A History of the Chrysler Corporation*. Detroit: Wayne State University Press, 2003.

Jackson, Kenneth T. *Crabgrass Frontier: The Suburbanization of the United States*. New York and Oxford: Oxford University Press, 1985.

Kirsch, David A. *The Electric Vehicle and the Burden of History*. New Brunswick, NJ: Rutgers University Press, 2000.

Lay, M. G. *Ways of the World: A History of the World's Roads and the Vehicles That Used Them*. New Brunswick, NJ: Rutgers University Press, 1992.

Lewis, David L., and Laurence Goldstein, eds. *The Automobile and American Culture*. Ann Arbor: University of Michigan Press, 1983.

Lewis, Tom. *Divided Highways: Building the Interstate Highways, Transforming American Life*. New York: Viking, 1997.

Liebs, Chester L. *Main Street to Miracle Mile: American Roadside Architecture*. Boston: Little, Brown and Co., 1985.

Lynd, Robert S., and Helen Merrell Lynd. *Middletown: A Study in Contemporary American Culture*. New York: Harcourt, Brace and Co., 1929.

Maxim, Hiram Percy. *Horseless Carriage Days*. New York and London: Harper & Brothers, 1937.

McShane, Clay. *Down the Asphalt Path: The Automobile and the American City*. New York: Columbia University Press, 1994.

Nader, Ralph. *Unsafe at Any Speed: The Designed-In Dangers of the American Automobile*. New York: Grossman, 1972.

National Bureau of Statistics of China. *China Statistical Yearbook*. N.p.: China Statistics Press, 2002.

Nevins, Allan, and Frank Ernest Hill. *Ford: The Times, the Man, the Company*. New York: Scribner, 1954.

———. *Ford, Expansion and Challenge, 1915–1933*. New York: Scribner, 1957.

Norbye, Jan P. *The Complete History of the German Car, 1886 to the Present*. New York: Portland House, 1987.

Patton, Phil. *Open Road: A Celebration of the American Highway*. New York: Simon and Schuster, 1986.

Posthumus, Cyril. *First Cars*. London: Phoebus, 1976.

Rae, John B. *The American Automobile: A Brief History*. Chicago: University of Chicago Press, 1965.

———. *The American Automobile Industry*. Boston: Twayne, 1984.

———. *The Road and the Car in American Life*. Cambridge, MA: MIT Press, 1971.

Rolt, L. T. C. *Motoring History*. London and New York: E. P. Dutton, 1964.

Rose, Mark. *Interstate: Express Highway Politics, 1941–1956*. Lawrence: Regents Press of Kansas, 1979.

Ross, Marc, and Tom Wenzel. "An Analysis of Traffic Deaths by Vehicle Type and Model." Washington, DC: American Council for an Energy-Efficient Economy, 2002.

Rubenstein, James M. *Making and Selling Cars: Innovation and Change in the U.S. Automotive Industry*. Baltimore and London: Johns Hopkins University Press, 2001.

Ruiz, Marco. *The Complete History of the Japanese Car, 1907 to the Present*. New York: Portland House, 1986.

Scarff, Virginal. *Taking the Wheel: Women and the Coming of the Motor Age*. New York: Free Press, 1991.

Sedgwick, Michael. *Cars of the Thirties and Forties*. New York: Crescent Books, 1992.

———. *Cars of the 50s and 60s*. New York: Beekman House, 1983.

Serrin, William. *The Company and the Union: The "Civilized Relationship" of the General Motors Corporation and the United Automobile Workers*. New York: Random House, 1974.

Sloan, Alfred P., Jr. *My Years with General Motors*. New York: Doubleday & Co., 1963.

South Coast Air Quality Management District web page, accessed at http://www.aqmd.gov/smog/o3trend.html.

Tubbs, D. B. *The Lancia Lambda*. Leatherhead, England: Profile Publications, n.d.

*2002 Market Data Book*. Detroit: *Automotive News*, 2002.

U.S. Census Bureau. *Statistical Abstract of the United States, 2001*. Washington, DC: U.S. Government Printing Office, 2002.

Walsh, Margaret. *Making Connections: The Long-Distance Bus Industry in the USA*. Aldershot, England: Ashgate, 2000.

Wik, Reynold M. *Henry Ford and Grass-Roots America*. Ann Arbor: University of Michigan Press, 1972.

Wilson, Paul C. *Chrome Dreams: Automobile Styling since 1893*. Radnor, PA: Chilton, n.d.

Wood, Jonathan. *Wheels of Misfortune: The Rise and Fall of the British Motor Industry.* London: Sidgwick & Jackson, 1988.

Wren, James A., and Genevieve J. Wren. *Motor Trucks of America.* Ann Arbor: University of Michigan Press, 1979.

Yates, Brock. *The Decline and Fall of the American Automobile Industry.* New York: Empire Books, 1983.

# Index

## About the Author

**RUDI VOLTI** is Professor of Sociology at Pitzer College in Claremont, California. His publications include *Society and Technological Change* (now in its 4th edition), *The Engineer in History* (now in its 2nd edition), *The Facts on File Encyclopedia of Science, Technology and Society*, and several articles and book reviews on the history of the automobile.